Ulrike Kindereit

Laser Voltage Probing for Electronic Devices

Ulrike Kindereit

Laser Voltage Probing for Electronic Devices

Detailed understanding of the physical signal origin and forecast about future scaling

Südwestdeutscher Verlag für Hochschulschriften

Impressum/Imprint (nur für Deutschland/only for Germany)
Bibliografische Information der Deutschen Nationalbibliothek: Die Deutsche Nationalbibliothek verzeichnet diese Publikation in der Deutschen Nationalbibliografie; detaillierte bibliografische Daten sind im Internet über http://dnb.d-nb.de abrufbar.
Alle in diesem Buch genannten Marken und Produktnamen unterliegen warenzeichen-, marken- oder patentrechtlichem Schutz bzw. sind Warenzeichen oder eingetragene Warenzeichen der jeweiligen Inhaber. Die Wiedergabe von Marken, Produktnamen, Gebrauchsnamen, Handelsnamen, Warenbezeichnungen u.s.w. in diesem Werk berechtigt auch ohne besondere Kennzeichnung nicht zu der Annahme, dass solche Namen im Sinne der Warenzeichen- und Markenschutzgesetzgebung als frei zu betrachten wären und daher von jedermann benutzt werden dürften.

Verlag: Südwestdeutscher Verlag für Hochschulschriften GmbH & Co. KG
Heinrich-Böcking-Str. 6-8, 66121 Saarbrücken, Deutschland
Telefon +49 681 37 20 271-1, Telefax +49 681 37 20 271-0
Email: info@svh-verlag.de

Approved by: Berlin, TU, Diss., 2008, D83

Herstellung in Deutschland:
Schaltungsdienst Lange o.H.G., Berlin
Books on Demand GmbH, Norderstedt
Reha GmbH, Saarbrücken
Amazon Distribution GmbH, Leipzig
ISBN: 978-3-8381-1059-2

Imprint (only for USA, GB)
Bibliographic information published by the Deutsche Nationalbibliothek: The Deutsche Nationalbibliothek lists this publication in the Deutsche Nationalbibliografie; detailed bibliographic data are available in the Internet at http://dnb.d-nb.de.
Any brand names and product names mentioned in this book are subject to trademark, brand or patent protection and are trademarks or registered trademarks of their respective holders. The use of brand names, product names, common names, trade names, product descriptions etc. even without a particular marking in this works is in no way to be construed to mean that such names may be regarded as unrestricted in respect of trademark and brand protection legislation and could thus be used by anyone.

Publisher: Südwestdeutscher Verlag für Hochschulschriften GmbH & Co. KG
Heinrich-Böcking-Str. 6-8, 66121 Saarbrücken, Germany
Phone +49 681 37 20 271-1, Fax +49 681 37 20 271-0
Email: info@svh-verlag.de

Printed in the U.S.A.
Printed in the U.K. by (see last page)
ISBN: 978-3-8381-1059-2

Copyright © 2012 by the author and Südwestdeutscher Verlag für Hochschulschriften GmbH & Co. KG and licensors
All rights reserved. Saarbrücken 2012

Contents

I. **Introduction**   5

II. **Theory**   8

1. **Propagation of light**   10
2. **Properties of laser light**   12
3. **Interaction of light and matter**   16
   - 3.1. Reflection, transmission and phase shift at interfaces; absorption in media   16
     - 3.1.1. σ-case 16
     - 3.1.2. π-case 17
     - 3.1.3. Phase shift at interfaces 18
     - 3.1.4. Reflectance and transmittance 19
   - 3.2. Interference effects 21
     - 3.2.1. Two-beam interference 21
     - 3.2.2. Multi-beam interference - Matrix formalism 22
4. **Relevant effects that influence the reflection from a device**   26
   - 4.1. Geometrical effects 26
   - 4.2. Wavelength dependent effects 26
     - 4.2.1. Optical absorption in intrinsic silicon 27
     - 4.2.2. Dispersion 27
   - 4.3. Temperature dependent effects 27
     - 4.3.1. Thermo-optic coefficient 29
   - 4.4. Effects depending on the electric field 30
     - 4.4.1. Franz-Keldysh effect or Electro-absorption 30
     - 4.4.2. Electro-refraction 32
     - 4.4.3. Kerr effect 32
     - 4.4.4. Pockels effect 33
   - 4.5. Free carrier effects or Plasma-optical effects 33
     - 4.5.1. Free carrier absorption 34
     - 4.5.2. Free carrier refraction 35

III. LVP setup and measurement methods   40

5. LVP setup and components   41
    5.1. General overview of LVP setups 41
    5.2. The LVP setup used for this work 41
         5.2.1. Interference of polarization states at the detector 43
         5.2.2. Laser wavelength (1064 nm versus 1319 nm) 44

6. Measurement methods and image acquisition   45
    6.1. Image acquisition 45
    6.2. LVP signal acquisition 45
         6.2.1. Time-domain measurement 46
         6.2.2. Frequency-domain measurement 46
         6.2.3. ppm-calculation 48

IV. Devices   52

7. Overview of the devices used   53
    7.1. Electrical characteristics of the devices 53
    7.2. Layouts of the devices 54
    7.3. Technology parameters 54
    7.4. Device modi 55
         7.4.1. Reverse biased diode 55
         7.4.2. Varactor in inversion 56
         7.4.3. FET, gate and drain simultaneously pulsed 59

V. Measurements   62

8. Detailed investigation of the images   64

9. Detailed investigation of modulation amplitude maps (MAM), modulation sign maps (MSM) and voltage sweeps (VS)   68
    9.1. Reverse biased drain diode ($V_G = V_S = V_W = GND$; V_D pulsed) 69
         9.1.1. 120 nm process technology 69
         9.1.2. 65 nm process technology 72
         9.1.3. Different device sizes 75
    9.2. Varactor in inversion ($V_D = V_S = V_W = GND$; V_G pulsed) 79
         9.2.1. 120 nm process technology 79
         9.2.2. 65 nm process technology 82
         9.2.3. Different device sizes 86
    9.3. FET, gate and drain pulsed ($V_S = V_W = GND$; $V_G = V_D$ pulsed simultaneously) 89
         9.3.1. 120 nm process technology 89

	9.3.2.	65 nm process technology	93
	9.3.3.	Different device sizes	98
9.4.	Signal contribution	101
	9.4.1.	Signal contribution - sign flips (MSMs) and signal level variations (MAMs) within the same area	101
	9.4.2.	Signal contribution - sign flip or signal amplitude variations in the VSs, signal level variations along the gate (gate signal)	104
	9.4.3.	FET signal contribution	104
	9.4.4.	Signal contribution summary	105
9.5.	Evaluation of the measurement methods	109

VI. Modeling 111

10. Active and passive signal contribution 112
10.1. Static part of the reflected light 112
10.2. Modulated part of the reflected light 112

11. Separate modeling of the interfaces 113
11.1. Effects in silicon 113

12. Simulations with the matrix formalism 119
12.1. Using the matrix formalism for reflectance simulations of the active areas of a FET 119
    12.1.1. Modeling of the drains of the FETs 119
    12.1.2. Modeling of the gates of the FETs 119
    12.1.3. Modeling of mobility for various carrier concentrations 121
    12.1.4. Calculation of the refractive index and the absorption coefficient . 122
12.2. Simulation results: n^+p-diode, reverse bias (NFET, drain) 122
    12.2.1. Influencing parameters 129
    12.2.2. Summary 130
12.3. Simulation results: p^+n-diode, reverse bias (PFET, drain) 131
    12.3.1. Influencing parameters 136
    12.3.2. Summary 137
12.4. Simulation results: NFET, gate 137
    12.4.1. Sub-threshold simulations 139
    12.4.2. Simulations above threshold 140
    12.4.3. Summary 141
12.5. Simulation results: PFET, gate 142
    12.5.1. Sub-threshold simulations 143
    12.5.2. Simulations above threshold 143
    12.5.3. Summary 144
12.6. Discussion and summary of the results 145
    12.6.1. Reverse biased diodes 145

    12.6.2. Varactors in inversion 145
    12.6.3. Limits of the Modeling 146

VII. Future prospects, summary and conclusions   148

13. Future Prospects   149
    13.1. Scaling of devices - future process technologies 149
    13.2. SOI 149

14. Summary and conclusions   152

VIII. Bibliography   155

IX. Acknowledgment   159

X. Appendix   162
    14.1. Free carrier absorption and refraction 163
    14.2. Calculation of the mobility 165
    14.3. Setup components - tool specification 166
        14.3.1. Oscilloscope 166
        14.3.2. Spectrum analyzer 166
        14.3.3. Function generator 167

Part I.
Introduction

Failure analysis of semiconductor devices is the technical field of investigating failures, which occurred during the manufacturing process of devices such as CMOSFETs (**c**omplimentary **m**etal **o**xide **s**emiconductor **f**ield **e**ffect **t**ransistors), memory and processor devices. The goal is to reduce development time by solving the problem early in the process. One important step in analyzing the failure is to measure the device activity. In the past, electrical micro-probing or e-beam probing were used to measure waveforms of devices from the front side. But these once invaluable tools become obsolete in recent years, due to the increasing complexity of the process technologies. To accommodate growing speed and integration density, multi-level wiring has been introduced. More and more of the chip area is covered with metalization, which obstructs access to internal measurement nodes [GSG93]. Probing signals from the backside of the devices (through the substrate), is the only access left to internal measurement nodes. The use of particle-based tools is limited for backside analysis, due to the extensive backside micro-structural modifications, which are necessary - e.g. FIB (focused ion beam), laser-etching etc..

Laser Voltage Probing (LVP) is the technical term for an all-optical laser-based technique that acquires waveforms through the silicon backside. Usually, LVP tools employ near infra-red (NIR) lasers. Since the thinned silicon bulk is partly transparent for NIR laser light, the backside only needs to be thinned (to about 100 μm) and polished. In conventionally used tools, the laser is pointed to one specific position and the semi-quantitative voltage level of the device at this position is extracted. The general signal generation develops as follows. The laser light gets reflected at the interfaces in the active areas of the device under test (DUT). The device voltage modulates the reflected light in the parts per million (ppm) regime. The setups consist of complex and cost-intensive detection schemes such as mode-locked (ML, or pulsed) lasers or phase detection methods (PID: phase interferometric detection), which make handling of the tools a quite challenging task. With high performance electronic equipment it is possible to detect the modulations in the reflection and thus evaluate the voltage level at the device.

Up to now, it is still not possible to extract exact voltage levels, because the signal strongly depends on focus, laser spot position, device geometry and details in the manufacturing process. In addition, the literature reveals several different assumptions about the interaction effects - of laser light and the active device - that are causing the modulations in the reflection of the laser light. If it is possible to determine the influencing effects, a reverse evaluation of the voltage levels might be achieved in the future.

The scope of this work is to evaluate a new, simplified, detection scheme, including new measurement methods at two different laser wavelengths (1319 nm and 1064 nm), and to analyze the physical background of LVP signal contribution in detail from the achieved results. In order to study the signal-to-voltage correlation that results from the interaction effects of laser light and the device activity, a model of LVP signal development will be built, which will verify the measurements. In addition, with the help of the measurements and the signal modeling, it will be possible to forecast the expected signal levels for future technologies and scaling.

This work presents measurements with a tool that employs a continuous wave (cw) laser. The laser wavelength can be chosen between 1319 nm (less resolution, but non-invasive) and 1064 nm (increased resolution, but possibly invasive, which means that the laser might influence the signal while the measurement takes place). In addition, new measurement methods - voltage sweeping and modulation sign and amplitude mapping - are introduced, which allow detailed investigations of signal-to-voltage information and signal tracking in an integrated circuit (IC).

The devices investigated for this work are test structure MOSFETs of two process technologies from Infineon Technologies AG. The measurements were performed on two device sizes: over-sized devices, to extract signals from gate and drain separately, and minimum-sized devices, in order to understand signal contribution of "real" devices, that are decreasing in size with growing speed and integration density.

A concise model of the modulated reflections from an active device is introduced - evaluating free carrier effects as the main effect of signal generation -, which allows simulations of the signal-to-voltage correlations of such signals. The results are in accord with the measurements.

The content of this work is structured as follows. First, all the interactions of the laser light with a semiconductor device will be outlined and evaluated, in order to determine the main effect, which causes the modulation in the reflected light due to the device activity (part II). This will be followed by a description of the setup, the measurement methods and the devices used (parts III and IV). The results of the performed measurements will be discussed in detail (part V). Part VI will show the modeling of LVP signals and present simulation results, which are then compared to the according measurements.

Part II.

Theory

An LVP tool measures the modulated part of the reflection of the laser light that is caused by the device activity. This part covers the theoretical background of laser light interaction with an IC. Properties of light in general and laser light in particular will be examined in chapter 1 and 2, followed by an outline of the interaction of light and matter (chapter 3); the last chapter deals with the relevant effects, which might occur in a switching device, while a laser beam is focused on its active areas.

1. Propagation of light

The classical theory predicted the **wave character** of light (light as an electro-magnetic wave that e.g. satisfies Maxwell's equations), later, in the early days of quantum mechanics, Einstein explained the photo-electric effect with the **particle character** of light (the light particle is called "photon"). Today we know that light can be understood as an electro-magnetic wave *or* a photon (**wave-particle dualism**), depending on the experimental conditions [1]. Accordingly, in the following, both - wave and particle - properties are used to explain the main characteristics of light.

The first three chapters describe light as an electro-magnetic wave, i.e. time- and space-dependent variations of coupled electric and magnetic fields. The **vector wave equations** for free space (vacuum) are expressions for these electro-magnetic waves.

For the electric field **E**:

$$\nabla^2 \mathbf{E} = \mu_0 \epsilon_0 \frac{\partial^2 \mathbf{E}}{\partial t^2}. \tag{1.1}$$

And the magnetic field **B**:

$$\nabla^2 \mathbf{B} = \mu_0 \epsilon_0 \frac{\partial^2 \mathbf{B}}{\partial t^2}. \tag{1.2}$$

The derivation of these equations from **Maxwell's equations** has been performed in the literature several times, but is not shown here. For further information see e.g. [Hec87]. In these equations, μ_0 is the permeability (the index 0 stands for "free space"), ϵ_0 is the electric permittivity and t is the time. In a homogeneous, isotropic dielectric medium μ_0 becomes $\mu_0 \mu_r = \mu$ (the index r stands for "relative") and ϵ_0 becomes $\epsilon_0 \epsilon_r = \epsilon$. The phase velocity in such a medium is defined as

$$v = \frac{1}{\sqrt{\mu \epsilon}}. \tag{1.3}$$

The ratio of the speed of an electro-magnetic wave in vacuum ($c_0 = \frac{1}{\sqrt{\mu_0 \epsilon_0}}$) to that in matter is known as the **absolute index of refraction n**:

$$n \equiv \frac{c_0}{v} = \sqrt{\frac{\mu \epsilon}{\mu_0 \epsilon_0}}. \tag{1.4}$$

---

[1] More precisely: In quantum theory, the location of an object (particle) is not defined, as long as it is not measured. Until then, one can only predict a probability to find the particle in one specific position. The probability in turn satisfies the wave equation, but this still does not mean that the particle itself can be understood as a wave.

The vector wave equations are derived under the condition that the media are non-conducting. In metal, for example, the time rate of change of **E** generates a voltage, currents circulate, and since the material is resistive, light is converted to heat - ergo **absorption** occurs (see [Hec87]). This leads to a **complex index of refraction** $\tilde{n}$:

$$\tilde{n} = n - ik. \tag{1.5}$$

Where n is the absolute index of refraction and k is the **extinction coefficient**, in which α is the **linear absorption coefficient** and λ is the wavelength:

$$k = \frac{\alpha \lambda}{4\pi}. \tag{1.6}$$

With the introduction of a complex index of refraction, the vector wave equations can be used for conducting materials (incl. semiconductors) as well, although they have been derived for non-conducting materials.

2. Properties of laser light

Plane waves are a quite simple solution of the vector wave equations. They are three-dimensional transversal electro-magnetic waves, which means that the electric and the magnetic field of the wave are perpendicular and both are perpendicular to the direction of propagation. The electric field component $\mathbf{E}(\mathbf{r},t)$ can be written as

$$\mathbf{E}(\mathbf{r},t) = \mathbf{E}_0 e^{i((\mathbf{kr}+\epsilon)\mp\omega t)}. \quad (2.1)$$

Where $\mathbf{E}_0$ is the amplitude of the electric field, $\mathbf{r} \equiv [x,y,z]$ is the position vector, here $\mathbf{k} \equiv [k_x, k_y, k_z]$ is the propagation vector and $k = \frac{2\pi}{\lambda}$ the propagation constant (or number). Physically only the real part of the electric field is of interest. The term ωt represents the variation in time (with $\omega = 2\pi f$; f: frequency of the wave). Further, in this equation, ϵ is the initial phase angle of the emitted wave (if two waves are emitted in phase, $\epsilon_1 = \epsilon_2$). Or alternatively, with $\Phi(r,\epsilon) = \mathbf{kr} + \epsilon$:

$$\mathbf{E}(\mathbf{r},t) = \mathbf{E}_0 e^{i(\Phi\mp\omega t)}. \quad (2.2)$$

The wavefronts of such waves are planes, which are determined by $\mathbf{kr} = const.$.

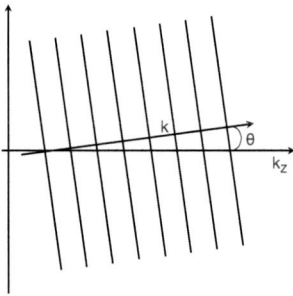

Figure 2.1.: Schematic of a plane wave with $\mathbf{kr} = const.$ traveling at a small angle θ to the optic axis, after [Sie86]

Plane waves are defined to be expanding to infinity, so laser light can not be described as a simple plane wave, because the beam radius, perpendicular to $\mathbf{k}$ (see figure 2.1), of laser light is finite.

Spherical waves are waves that have an imaginary point source, which expands radially uniformly in all directions. The wavefronts here are concentric spheres. Their electrical field component can be written as

$$\mathbf{E}(\rho, t) = \frac{\mathbf{E}_0}{\rho} e^{i(\mathbf{k}\rho \mp \omega t)}, \qquad (2.3)$$

where $\rho = \sqrt{x^2 + y^2 + z^2}$ is the radius from the imaginary source. However, laser light has only one direction of propagation, so again, it can not be described with the model of a spherical wave.

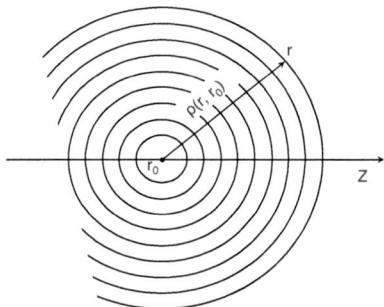

Figure 2.2.: Schematic of a general spherical wave, after [Sie86]

A better way of describing **laser light** is to observe a plane wave that traverses a slit that is determining the wave in x- and y-direction. The result of such a wave is shown in figure 2.3.

The intensity profile of a laser beam versus its radius is following the **Gaussian distribution**, see figure 2.4.

Here, the waist of a Gaussian beam w_0 is defined as the radius from the center of the beam to the point, where the intensity of the beam is decreased to e^{-1}. The power transmission of the Gaussian beam through a circular aperture with the radius $a = w_0$ is $\approx 86\%$ (compare to [Sie86]). Figure number 2.5 shows a collimated waist region of such a Gaussian beam. The "spot size" of the beam is then $2w_0$. The beam expands as it propagates from the waist region. The distance, which the beam travels from the waist until the beam diameter increases by $\sqrt{2}$, or until the beam area doubles, is given by

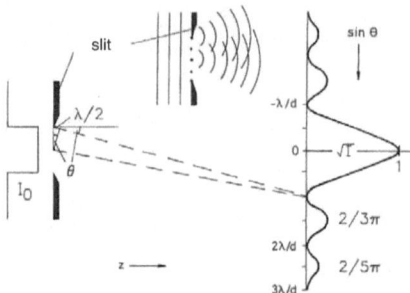

Figure 2.3.: Huyghen's principle: diffraction of a plane wave at a slit with the width d [EE90]

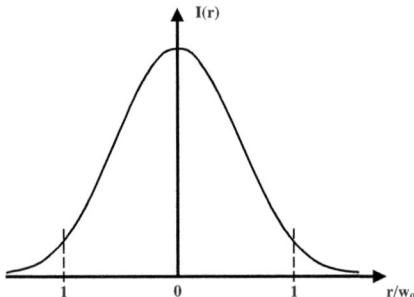

Figure 2.4.: Gaussian distribution - intensity I versus radius r of a circular Gaussian beam, after [Sie86]

$$z = z_R \equiv \frac{\pi w_0^2}{\lambda}, \qquad (2.4)$$

the **Rayleigh range**, which divides the near-field (Fresnel regions) from the far-field (Fraunhofer regions).

The "depth of focus" or the **confocal parameter b** is given by $b = 2z_R$. In the center (z=0), the wavefronts are plane. To simplify matters, it will be assumed that within the confocal parameter, laser light can be understood as a plane wave and will be treated

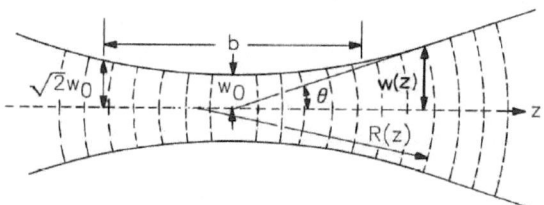

Figure 2.5.: Collimated waist region of a Gaussian beam; the diagram shows the propagation of a Gaussian beam: changes of the beam radius w(z) and of the radius of curvature R(z) of the wave fronts [EE90]

like that in the following paragraphs.

3. Interaction of light and matter

The following sections give an overview of the interactions of matter and a light wave - a plane wave, as which laser light has been approximated in the previous chapter. First, a general description of reflection, transmission and absorption will be given, including the phase shift at interfaces, then interference effects will be discussed in detail. The dependency of the reflection on the device activity will be investigated in chapter 4.

3.1. Reflection, transmission and phase shift at interfaces; absorption in media

When a light wave is incident on a plane surface, the light can be reflected and refracted. Neglecting multi-reflections at other interfaces for now, the refracted part of the wave can either be transmitted through the medium or it will be absorbed by the medium. The following paragraphs will deal with the Fresnel equations - which describe, whether there is a phase shift between the incident wave and the reflected / transmitted wave - and the reflectance and transmittance - which determine the ratio of the reflected / transmitted power to the incident power in percent.

A complete derivation of the general Fresnel equations can e.g. be found in [Hec87]. However, here only the simplified equations for non-magnetic materials, where $\mu_r = 1$, so $\mu = \mu_0$, for all materials used ($\mu_i = \mu_t = \mu_0$; indices: i-medium of incidence, t-medium of transmission) are shown. The Fresnel equations differ for the two possible orientations between the electrical field of the laser beam and the plane of incidence (see figure number 3.1):

3.1.1. σ-case

If the electric field is oriented perpendicular to the incident plane, the **amplitude reflection coefficient** is calculable as follows:

$$\rho_\sigma = \left(\frac{E_{0r}}{E_{0i}}\right) = \frac{\tilde{n}_i cos(\theta_i) - \tilde{n}_t cos(\theta_t)}{\tilde{n}_i cos(\theta_i) + \tilde{n}_t cos(\theta_t)}. \tag{3.1}$$

For the case that the laser beam is perpendicular to the surface of the device (normal incidence), the angle θ_i is zero, which simplifies the equation even further ($\theta_i = \theta_t$):

$$\rho_\sigma = \frac{\tilde{n}_i - \tilde{n}_t}{\tilde{n}_i + \tilde{n}_t}. \tag{3.2}$$

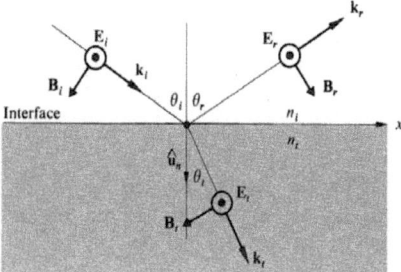

Figure 3.1.: Definition of the terms used here: plane of incidence and orientation of the electric field to it, shown here: a wave whose electric field is normal to the plane of incidence [Hec87]

The **amplitude transmission coefficient** for this case is defined as

$$\tau_\sigma = \left(\frac{E_{0t}}{E_{0i}}\right) = \frac{2\tilde{n}_i cos(\theta_i)}{\tilde{n}_i cos(\theta_i) + \tilde{n}_t cos(\theta_t)}. \tag{3.3}$$

And again, with the assumption that the laser beam is oriented perpendicular to the incident plane, this simplifies to

$$\tau_\sigma = \frac{2\tilde{n}_i}{\tilde{n}_i + \tilde{n}_t}. \tag{3.4}$$

In this case, $\tau_\sigma + (-\rho_\sigma) = 1$ (this is true for all incident angles).

3.1.2. π-case

If the electric field is oriented parallel to the incident plane, the **amplitude reflection coefficient** for normal incidence is

$$\rho_\pi = \frac{\tilde{n}_t - \tilde{n}_i}{\tilde{n}_t + \tilde{n}_i} = -\rho_\sigma. \tag{3.5}$$

The negative sign in this equation compared to equation 3.2 means that a 180° phase shift of the electric field component occurs, when the first (incident) medium has a higher index of refraction as the second medium, compare to section 3.1.3 (note that this is only true for normal incidence).

The **amplitude transmission coefficient** for this case is (for normal incidence)

$$\tau_\pi = \frac{2\tilde{n}_i}{\tilde{n}_i + \tilde{n}_t}. \tag{3.6}$$

Here $\tau_\pi + \rho_\pi = 1$ (only true for the assumption of normal incidence).

3.1.3. Phase shift at interfaces

Figure number 3.2 shows the result of the amplitude reflection coefficients for reflection at an interface. The two media consist of the indices of refraction n_1=3.4975 and n_2, where n_2 is varied within 3.495 and 3.5.

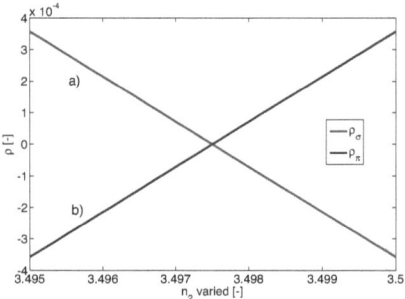

Figure 3.2.: Amplitude reflection coefficients for reflection at a single interface with n_1=3.4975 and n_2 varied. a) σ-case (for the electric field component perpendicular to the incident plane) b) π-case (electric field component parallel to the incident plane)

In this graph, two cases are shown: ρ_σ is the reflection coefficient for the electric field component oriented perpendicular to the incident plane; ρ_π is the reflection coefficient for a parallel orientation of the electric field. The meaning of the negative sign of the reflection coefficient is that a phase shift of π radians takes place, so the component of the electric field undergoes a 180° phase shift. As can be found in the graph, for the $\sigma(\pi)$-case this always happens, when the first medium has a lower (higher) index of refraction then the second medium (**external reflection** for $n_i < n_t$ ($n_i > n_t$)). The phase of the electric field stays the same, as long as the index of refraction of the first medium is higher (lower) or more (less) dense: $n_i > n_t$ ($n_i < n_t$), which is called **internal reflection**. Note that this is only true for light that traverses the interface from medium 1 to 2. If there was light being reflected at interfaces beyond the structure, so that the light traverses in the opposite direction - from medium 2 to 1 - of course, the relations change, due to the opposite ratio of the indices of refraction.

Figure number 3.3 shows the amplitude transmission coefficients for the two cases and the structure described above.

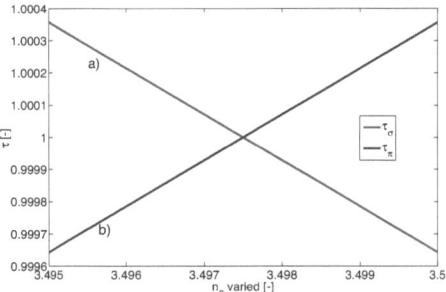

Figure 3.3.: Amplitude transmission coefficients for reflection at a single interface with $n_1=3.4975$ and n_2 varied. a) σ-case (for the electric field component perpendicular to the incident plane) b) π-case (electric field component parallel to the incident plane)

The transmission coefficient is always positive, which means that - regardless of the indices of refraction of the two media - there is no phase shift in the component of the electric field being transmitted through the interface compared to the incident component.

3.1.4. Reflectance and transmittance

The **reflectance R** (ratio of the reflected power to the incident power) and the **transmittance T** (ratio of the transmitted power to the incident power) are defined as follows (for normal incidence) ($R + T = 1$):

$$R = \left| \frac{E_{0r}}{E_{0i}} \right|^2 = |\rho|^2 , \tag{3.7}$$

$$T = \frac{\tilde{n}_t}{\tilde{n}_i} \left| \frac{E_{0t}}{E_{0i}} \right|^2 = \frac{\tilde{n}_t}{\tilde{n}_i} |\tau|^2 . \tag{3.8}$$

Figure number 3.4 and 3.5 show the reflectance and the transmittance for the situation described in section 3.1.3. Since both - reflectance and transmittance - are calculable with the according amplitude coefficients squared, the result is always positive and thus the results for the two different orientations of the electric field to the incident plane are the same.

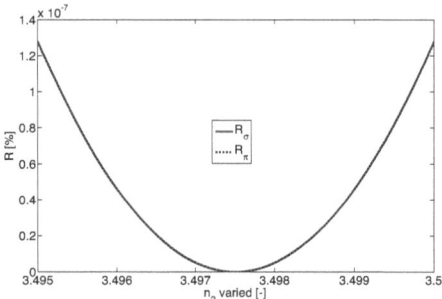

Figure 3.4.: Reflectance at a single interface with $n_1=3.4975$ and n_2 varied. σ- and π-case results are equal.

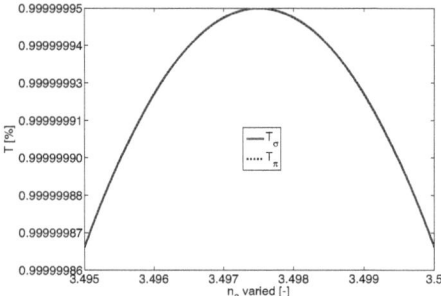

Figure 3.5.: Transmittance through a single interface with $n_1=3.4975$ and n_2 varied. σ- and π-case results are equal.

The combination of the amplitude reflection coefficients and the reflectance of an interface describes, what happens with the reflected component of the electric field of the incident beam: the reflectance provides the ratio of the reflected power to the incident power in percent and the amplitude reflection coefficient points out, whether there is a phase shift between the incident wave and the reflected wave.

3.2. Interference effects

In the paragraphs above, the properties of only one interface were described. However, in a real device, there are many different interfaces (compare to table 11.2) with the according properties such as indices of refraction, reflection coefficients and phase shifts between the incident wave and the reflected wave. When all these parts of the reflected beam interfere, the overall reflection (LVP signal, compare to section 4) will change in amplitude and phase. The following paragraphs will examine, how the parts of the beam that gets reflected will interfere.

Suppose that there are N overlapping waves having the same frequency and traveling in the positive x-direction, the resulting wave is given by superposition (compare to equation number 2.2):

$$\mathbf{E} = \left[\sum_{j=1}^{N} \mathbf{E}_{0j} e^{i\Phi_j}\right] e^{i\omega t}. \tag{3.9}$$

The composite wave is harmonic and of the same frequency as the constituents, but amplitude and phase are different.

$\mathbf{E}_0 e^{i\Phi} = \sum_{j=1}^{N} \mathbf{E}_{0j} e^{i\Phi_j}$ is the complex amplitude of the composite wave and $\mathbf{E}_0$ can be calculated from $\mathbf{E}_0 = \sqrt{\left|\mathbf{E}_0^2\right|} = \sqrt{(\mathbf{E}_0 e^{i\Phi})(\mathbf{E}_0 e^{i\Phi})^*}$.

3.2.1. Two-beam interference

For two-beam interference (N=2) the composite wave is calculable as follows:

$$\begin{aligned}\mathbf{E}_0 &= \sqrt{(\mathbf{E}_{01} e^{i\Phi_1} + \mathbf{E}_{02} e^{i\Phi_2})(\mathbf{E}_{01} e^{-i\Phi_1} + \mathbf{E}_{02} e^{-i\Phi_2})} \\ &= \sqrt{\mathbf{E}_{01}^2 + \mathbf{E}_{02}^2 + \mathbf{E}_{01}\mathbf{E}_{02}\left[e^{i(\Phi_1-\Phi_2)} + e^{-i(\Phi_1-\Phi_2)}\right]},\end{aligned} \tag{3.10}$$

or

$$\mathbf{E}_0 = \sqrt{\mathbf{E}_{01}^2 + \mathbf{E}_{02}^2 + 2\mathbf{E}_{01}\mathbf{E}_{02} cos(\Phi_1 - \Phi_2)}. \tag{3.11}$$

So the result is not simply the sum of the components, but there is an additional contribution $2\mathbf{E}_{01}\mathbf{E}_{02} cos(\Phi_1 - \Phi_2)$, known as the **interference term**. The crucial factor is the difference of the phase between the two interfering waves $\mathbf{E}_{01}$ and $\mathbf{E}_{02}$ $\delta \equiv \Phi_1 - \Phi_2$. When $\delta = 0, \pm 2\pi, \pm 4\pi, \ldots$ the resultant amplitude is a maximum (**constructive interference**), whereas $\delta = \pm\pi, \pm 3\pi, \ldots$ yields a minimum (**destructive interference**). Since $\Phi(r,\epsilon) = \mathbf{kr} + \epsilon$ the **phase difference / shift** may arise from a **difference in path length** traversed by the two waves, as well as a difference in the initial phase angle. If the two waves are initially in phase ($\epsilon_1 - \epsilon_2 = 0$)

$$\delta = \frac{2\pi}{\lambda_0}\tilde{n}(x_1 - x_2) \tag{3.12}$$

(for x-direction), with $\tilde{n}$ the index of refraction of the medium the wave traverses. Then the quantity $\tilde{n}(x_1-x_2)$ is known as the **optical path difference**. If $\epsilon_1 - \epsilon_2 = const.$ the waves are said to be **coherent**. The phase of the composite wave is calculable as follows: $\Phi = arctan\left(\frac{Im(E)}{Re(E)}\right)$. From equations number 3.11, 3.12 and 3.7 a proportionality of the reflectance resulting from two interfering waves can be derived:

$$R \propto \mathbf{E}_{01}^2 + \mathbf{E}_{02}^2 + 2\mathbf{E}_{01}\mathbf{E}_{02}cos(\delta). \tag{3.13}$$

If the amplitudes of the two waves were the same, $\mathbf{E}_{01} = \mathbf{E}_{02}$, this lead to:

$$R \propto \left\{ \begin{array}{c} 2R_{01}(\delta = 0, 2\pi, ...) \\ R_{01}(\delta = \frac{\pi}{2}, \frac{3\pi}{2}, ...) \\ 0(\delta = \pi, 3\pi, ...) \\ x \cdot R_{01}(\delta = \frac{2\pi}{\lambda_0}\tilde{n}(x_1 - x_2)) \end{array} \right\}. \tag{3.14}$$

This equation shows that the reflectance in such a situation can vary between 0 and $2R_{01}$ depending on the phase shifts between the two waves, which can arise from a difference in path length or from the reflection off an interface that shifts the phase by π due to the properties of the index of refraction (see π- and σ-case in section 3.1). In opposition to the situation just described ($\mathbf{E}_{01} = \mathbf{E}_{02}$), it is possible that the amplitudes are not the same due to the properties of the interfaces from which the waves are reflected. This also effects the strength of the reflectance.

3.2.2. Multi-beam interference - Matrix formalism

Until now, multiple reflections at interfaces and the resulting interferences were neglected. If the reflections at the interfaces are large, these reflections need to be taken into account, because the different parts of the beam interfere and are all contributing to the overall reflectance. All the reflections of the beam are partly phase shifted in comparison to the others and the amplitude is decreased due to the according product of reflection and transmission coefficients. If the structure of interest is only one layer (e.g. a plate), it is possible to calculate the reflectance of the plate with the multi-reflections included by hand, but if the medium consists of more layers, the calculation gets fairly complex. In the following paragraph, a formalism to calculate the reflectance (for normal incidence) of such a structure is explained [KF88].

Figure 3.6 shows a structure with multi-reflections. The angle θ shown in the graph is $0°$ for normal incidence. The interference effects are caused by the difference in path length and the properties of the interfaces. Figures number 3.7 and 3.8 show the notation for the matrix calculations described in the following. The light enters the system from the left. Two arbitrary adjacent layers are labeled i and j. The optical electrical field in each layer consists of two components: for the layer j E_{rj} travels to the right and E_{lj}

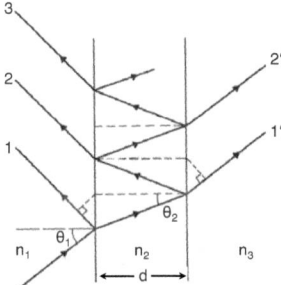

Figure 3.6.: Beam propagation for multi-beam interference, after [KF88]

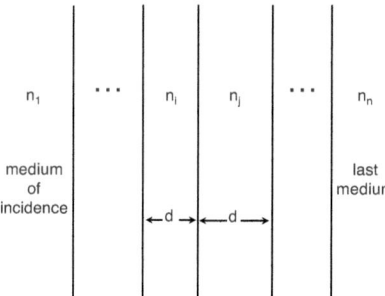

Figure 3.7.: Matrix formalism: Terms for the layers and their properties, after [KF88]

travels to the left. Each layer has two sides; the fields at the left side are not labeled in addition, the fields at the right side are labeled with a dash. Matrices at each point in the system can now be declared. For the fields at the right side of layer i the matrix can be written as

$$\mathbf{E}'_i \equiv \begin{pmatrix} \mathbf{E}'_{li} \\ \mathbf{E}'_{ri} \end{pmatrix}. \tag{3.15}$$

Analogously for layer j:

$$\mathbf{E}'_j \equiv \begin{pmatrix} \mathbf{E}'_{lj} \\ \mathbf{E}'_{rj} \end{pmatrix}, \tag{3.16}$$

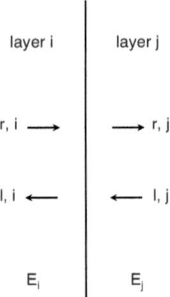

Figure 3.8.: Matrix formalism: Terms for the fields at both sides of an interface between layer i and j, the waves travel *from* the interface or *to* the interface, after [KF88]

and at the left side of this layer:

$$\mathbf{E}_j \equiv \begin{pmatrix} \mathbf{E}_{lj} \\ \mathbf{E}_{rj} \end{pmatrix}. \tag{3.17}$$

When the wave hits the interface of layer i and j, the according amplitude reflection and transmission coefficients are ρ_{ij} and τ_{ij} (compare to equation number 3.2 et sqq.). The **transfer matrix H** for this interface is defined as

$$\mathbf{H}_{ij} \equiv \frac{1}{\tau_{ij}} \begin{pmatrix} 1 & \rho_{ij} \\ \rho_{ij} & 1 \end{pmatrix}. \tag{3.18}$$

So for example the fields at the right side of layer i can be described with

$$\mathbf{E}'_i = \mathbf{H}_{ij} \mathbf{E}_j. \tag{3.19}$$

The propagation of light through layer j is defined by its **propagation matrix L**.

$$\mathbf{L}_j \equiv \begin{pmatrix} e^{-i\delta_j} & 0 \\ 0 & e^{i\delta_j} \end{pmatrix} \tag{3.20}$$

In this equation δ is the difference in path length or **phase factor**, which has already been defined in equation 3.12. The fields at the left side of layer j can now be calculated with

$$\mathbf{E}_j = \mathbf{L}_j \mathbf{E}'_j. \tag{3.21}$$

One boundary condition needs to be defined: in the last medium, there should be no wave which travels to the left side, so

$$\mathbf{E}_N = \begin{pmatrix} 0 \\ \mathbf{E}_{rN} \end{pmatrix}. \tag{3.22}$$

Exercising the matrix multiplications above will show that the matrix formalism is a concise way of calculating the electric fields at any point in the system (see [KF88] for more explanations). Further the entire system can be described with the matrix multiplication. Equation number 3.23 shows how the **system matrix S** can be expanded easily for N media:

$$\mathbf{H}_{12}\mathbf{L}_2\ldots\mathbf{L}_{N-1}\mathbf{H}_{N-1,N} \equiv \mathbf{S}_{1N} = \begin{pmatrix} S_{11} & S_{12} \\ S_{21} & S_{22} \end{pmatrix}. \tag{3.23}$$

The overall amplitude reflection and transmission coefficients (for π- or σ-case, depending which orientation was used for the declaration of equation number 3.18) are calculable as follows:

$$\rho \equiv \frac{\mathbf{E}'_{l1}}{\mathbf{E}'_{r1}} = \frac{\mathbf{S}_{12}}{\mathbf{S}_{22}}, \tag{3.24}$$

$$\tau \equiv \frac{\mathbf{E}_{rN}}{\mathbf{E}'_{r1}} = \frac{1}{\mathbf{S}_{22}}. \tag{3.25}$$

And with equation number 3.7 the overall reflectance (LVP signal, see section 4) of the system can be extracted:

$$R = |\rho|^2. \tag{3.26}$$

4. Relevant effects that influence the reflection from a device

As described in chapter 3, the overall reflectance of a structure is calculable with the index of refraction, the extinction coefficient (see equations 3.2-3.7) - i.e. absorption coefficient - and the thicknesses of the according layers (see equation 3.12). Most of the layers that can be found in a MOSFET structure have properties, which are independent of the voltage that is applied to the DUT. However, the optical properties of silicon depend on the laser wavelength - in LVP tools, NIR-lasers are used - , the temperature, the electric field and the free carrier concentration. In addition to these effects, in the active area of a MOSFET, the optical properties change with the variations in the electrical properties due to the applied device voltage, as well. Not only the thicknesses of the space charge regions (SCR) and the inversion channels are modulated, but the index of refraction and the extinction coefficient vary due to the device action, too. This chapter outlines the effects, which change the optical properties as described above. The resulting influence on the reflectance will be discussed in detail in part VI. Chapter 6 will describe, how the two parts of the reflection (the static part and the modulated part) are used to produce an image and the LVP signal.

4.1. Geometrical effects

In the static condition of a device, the index of refraction and the extinction coefficient at a point (x, y) are depending on the position of the laser beam. For example, in a large FET, drain and gate can be distinguished from each other, so the laser can be pointed to the drain or the gate of the structure. The optical parameters differ for both positions: the refractive indices, the extinction coefficients and the thicknesses of the layers are distinctive, which in turn influences the reflectance for both positions. If the device is stimulated electrically, the parameters of the layers - such as SCR and inversion channels - will change with the voltage and thus influence the LVP signal, too.

4.2. Wavelength dependent effects

The following paragraph reviews the absorption coefficient and the index of refraction for intrinsic silicon depending on the wavelength. The other effects show a wavelength-sensitivity in addition, but the according dependencies are described in the paragraphs separately.

4.2.1. Optical absorption in intrinsic silicon

When photons are absorbed by the semiconductor, generation of electron-hole pairs occurs, if the conservation of momentum and energy are satisfied: The energy and the momentum of the photons are transferred to the electrons and the phonons (quasi particle that describes lattice vibrations, energy in the range of thermal energy) during the interaction. The process of generating electron-hole pairs (inter-band absorption) can occur, if the total amount of energy of the photon and the assisting phonon is equal to or higher than the band-gap energy of the semiconductor. If the photon energy is higher than the band-gap energy, the energy difference is finally transformed into heat energy (phonons) until the electron-hole pair reaches minimum energy (i. e. the band-gap energy). If the photon energy is lower than the band-gap energy, electron-hole pair generation will usually *not* take place, but the absorbed energy will be transformed into thermal energy (intra-band absorption causes lattice vibrations, phonons). There are also phonon-assisted generation processes for sub-band-gap energies, but they are less probable.

The band-gap energy of intrinsic silicon is 1.12 eV. The energy of a photon of the laser light is calculable by

$$E = \frac{hc}{\lambda}, \tag{4.1}$$

where h is Planck's constant. So for wavelengths lower than 1107 nm (energy higher than 1.12 eV) electron-hole pair generation is likely to occur. The absorption coefficient depending on the wavelength (photon energy) is illustrated in figure 4.1 (figure 4.2). Graph 4.2 clearly shows the relatively steep slope around 1.12 eV, the band-gap energy of silicon. For indirect semiconductors (Si, Ge) the conservation of momentum can only be satisfied with the interaction of phonons, which makes the process less probable in comparison to direct semiconductors. This is the reason, why the slope of the absorption coefficient versus the energy of the photons for indirect semiconductors is not as steep as the slope of direct semiconductors (GaAs, compare to [WE94]).

4.2.2. Dispersion

The dependency of the **index of refraction** on the **wavelength** is called dispersion. Figure number 4.3 shows the dispersion behavior of intrinsic silicon (Si) at 300 K [GK95].

4.3. Temperature dependent effects

As discussed in section 4.2.1, besides electron-hole pair generation, the photon energy of the laser light will be transformed into thermal energy (phonons). The energy that is coupled into the device, while the measurement is performed (laser scanning across

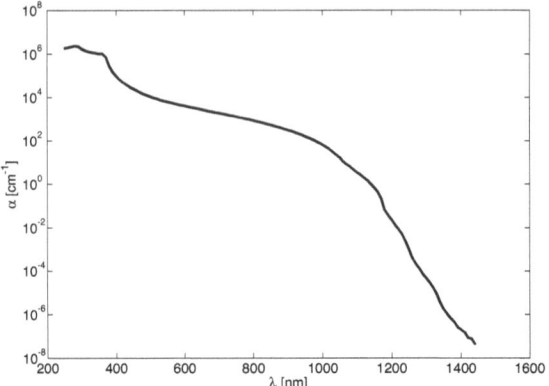

Figure 4.1.: Absorption coefficient (semi-logarithmic scale) of intrinsic silicon depending on the wavelength [GK95].

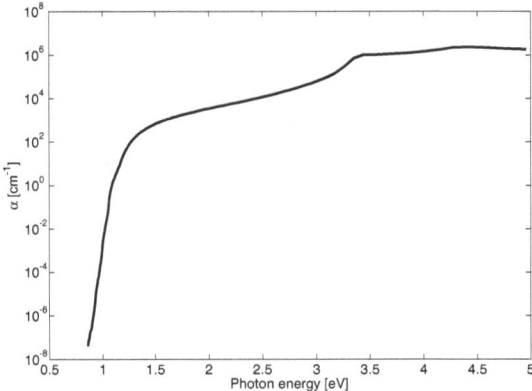

Figure 4.2.: The absorption coefficient (semi-logarithmic scale) of intrinsic silicon depending on the photon energy [GK95].

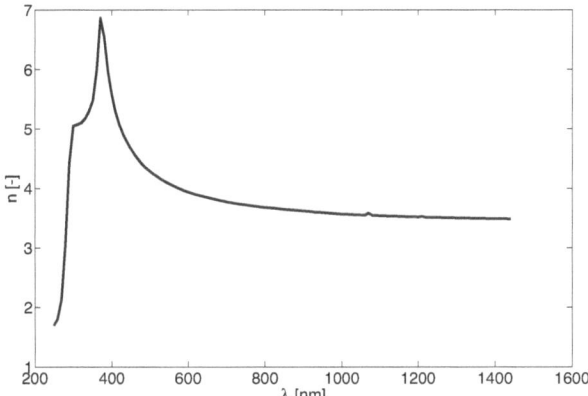

Figure 4.3.: Dispersion: index of refraction depending on the wavelength, graph shows data for intrinsic Si at 300 K [GK95].

the structure or pointed statically to one position), can be calculated by the following equation:

$$E = Pt. \tag{4.2}$$

Here, P is the laser power and t is the time the sample is exposed to the laser beam. To prevent the device from the influence of the thermal energy or even from destruction, either the laser power or the scanning time can be reduced (the higher the scanning speed, the lower the energy impact). For a good signal-to-noise ratio, it is necessary to decrease the scanning speed, so the factor that is determining the energy impact on the device is the laser power. Other LVP-setups employ pulsed lasers (not used for this work, compare to chapter 5), which allow the use of higher laser powers: with a shorter exposure time the energy impact can be controlled.

4.3.1. Thermo-optic coefficient

The thermo-optic coefficient is defined as $\frac{\Delta n}{\Delta T}$ [1/K]. The thermo-optic coefficients of silicon for various wavelengths and temperatures were measured by Frey et al., the results can be found in figure number 4.4 [FLM06]. To determine the thermo-optic coefficient, the peak temperature variations, caused by the laser, need to be evaluated (for the evaluation of the thermo-optic coefficient according to the lasers employed in this work see section 11.1).

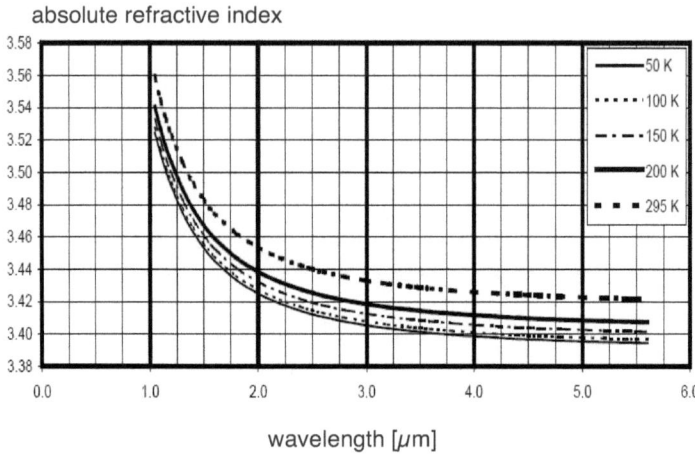

Figure 4.4.: Measurement of the absolute index of refraction of silicon as a function of wavelength for selected temperatures [FLM06]

4.4. Effects depending on the electric field

There are several active electric fields in a FET - e.g. in the SCR of the drain-to-well junction and underneath the gate (varactor). The strengths of both electric fields (E_{appl}) alter with the voltage that is applied. Since magnetic fields can be neglected in silicon devices, the magneto-optic effects (such as the Faraday effect, which rotates the plane of polarization) are not covered here.

4.4.1. Franz-Keldysh effect or Electro-absorption

The Franz-Keldysh effect is a photon assisted tunnel effect, which alters the fundamental optical absorption spectrum (see section 4.2.1) of a semiconductor in the presence of a high electric field. In 1957 / 1958, this effect has been described by W. Franz [Fra58] and L. V. Keldysh [Kel58] independently for the first time (for further information see [Kel64], [Wil60], [Pan71]). The change in the fundamental absorption is caused by a reduction of the effective band-gap energy. As described in 4.2.1, a photon with an energy that is lower than the band-gap energy – in general – will not generate an electron-hole pair (sharp edge in the absorption spectrum). A valence electron that has a lower energy than the band-gap energy could only get into the conduction band, if it *tunnels* an energy barrier. The height of this barrier is depending on the band-gap energy and the photon energy. The thickness of the barrier is depending on the electric field (the higher the field the thinner the barrier). If a high electric field, in the range of $10^5 \frac{V}{cm}$, is applied, the edges of the valence and conduction band will bend and thus decrease the thickness of the energy barrier, which will in turn increase the probability of *tunneling* [Kel57]. So with an increasing electric field, the sharp edge in the absorption spectrum is oblate for photon energies lower than the band-gap energy, i.e. the absorption coefficient is increased.

The electro-absorption spectrum has been measured by Wendland and Chester. Figure 4.5 shows the electro-absorption spectrum of silicon for various strengths of an electric field [WC65].

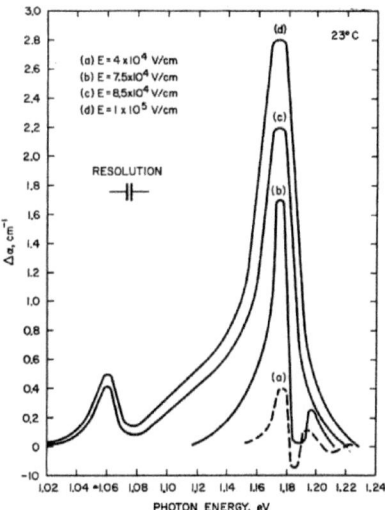

Figure 4.5.: Electro-absorption spectrum of silicon for various strengths of an electric field [WC65]

The graph shows two major electro-absorption peaks at photon energies of 1.175 eV (corresponding to 1055 nm) and at 1.06 eV (and 1170 nm laser wavelength).

4.4.2. Electro-refraction

Electro-refraction is the term for the dependency of the index of refraction on the electric field. Using the Kramers-Kronig relation (a correlation of absorption and dispersion), Soref and Bennett calculated the electro-refraction from the data shown in figure 4.5 (electro-absorption) - the result is shown in figure 4.6.

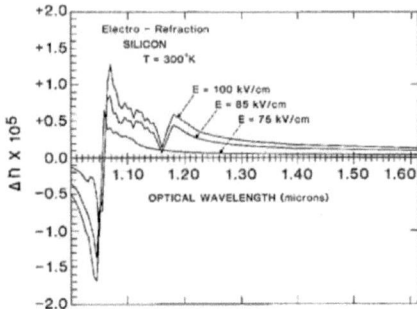

Figure 4.6.: Electro-refraction spectrum in silicon [SB87]

Note that Δn is positive for wavelengths higher than 1050 nm and negative for wavelengths lower than 1050 nm and is a strong function of wavelength. Further the authors found an increase rate of $\frac{\Delta n}{\Delta E}$ slightly stronger than E^2 ($E \approx (aE^2)$) and they predict a polarization dependence of electro-refraction: it will be 2x stronger for $E_{opt} \parallel E_{appl}$ than for $E_{opt} \perp E_{appl}$.

4.4.3. Kerr effect

Another effect that describes the dependency of the index of refraction on the electric field is the Kerr effect [HA79]. The applied electric field causes an orientation of the charge carriers in the material (crystals and liquids), which in turn causes birefringence. The Kerr effect depends on the orientation and polarization of the incident beam and the orientation of the applied electric field.

Soref and Bennett estimated the Kerr effect in silicon using the anharmonic oscillator model [SB87]:
$$\Delta n = -3q^2(n^2 - 1)E^2/2nm^2\omega_0^4 x^2, \tag{4.3}$$
which is proportional to the electric field E squared and independent of the wavelength. In this equation q is the electronic charge, n is the unperturbed index of refraction of silicon, m the effective mass, ω_0 the oscillator resonance frequency and x the average

oscillator displacement.

Figure number 4.7 shows the result of this approximation for λ=1300 nm ($\omega_0 = 2\pi \cdot 10^{15}$ rad/s; $x = 10^{-9}$ m). The authors describe an uncertainty about the sign of the Kerr effect: the anharmonic oscillator model predicts a negative sign, where an experiment suggested a positive sign.

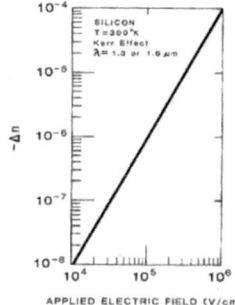

Figure 4.7.: Kerr effect in silicon versus E as determined from the anharmonic oscillator model [SB87]

4.4.4. Pockels effect

The Pockels effect describes - similarly to the Kerr effect - birefringence in crystals caused by an electric field. In contrary to the Kerr effect, here the dependency of the refractive index on the applied electric field is linear. This effect is present only in materials that do not have a symmetrical crystal structure (e.g. Gallium-Arsenide). Since silicon is a symmetrical crystal, this effect is negligible.

4.5. Free carrier effects or Plasma-optical effects

Free carrier effects describe the change in the index of refraction and the absorption coefficient in silicon depending on the density of free carriers. In addition to the free carriers that are present in a semiconductor device in static condition, free carriers can also be injected ($+\Delta N$) depending on the voltage that is applied to the device. One example for the presence of free carriers in a FET is the inversion channel.

4.5.1. Free carrier absorption

The changes in the absorption coefficient (and the index of refraction) caused by free carriers can be derived from the classical theory of dispersion in dielectrics. A complete derivation of equations number 4.4 and 4.5 is shown in the appendix (14.1). The calculations predict changes in the absorption coefficient according to:

$$\Delta \alpha = \frac{\lambda^2 q^3}{4\pi^2 c_0^3 \epsilon_0} \cdot \frac{\Delta N}{nm^2 \mu}. \tag{4.4}$$

In the above, $\Delta \alpha$ is the change in the absorption coefficient (positive), μ the mobility and ΔN the change in charge carrier densities. Here $n = n_0 + \Delta n$, where n_0 is the index of refraction with taking dispersion into account (compare to section 4.2.2) and Δn the change in the index of refraction due to free carriers calculable from equation 4.5.

Soref and Bennett also investigated the free carrier effects in silicon (free carrier absorption and refraction - for the latter see section 4.5.2). The authors collected absorption spectra data based on free carrier effects from the literature and presented the combined absorption spectra from 0.001 up to 2.8 eV for free electrons and holes (see figure 4.8 and 4.9 [SB87]).

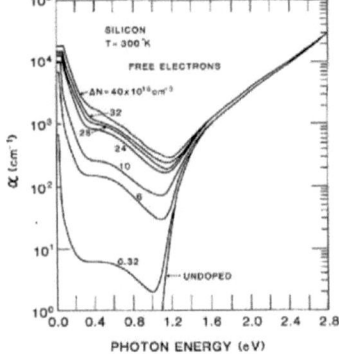

Figure 4.8.: Optical absorption spectra of silicon showing the influence of various concentrations of free electrons [SB87]

In graphs number 4.10 and 4.11, the authors plotted the experimental results and the theoretically achieved data versus the free carrier concentration. The results reveal that

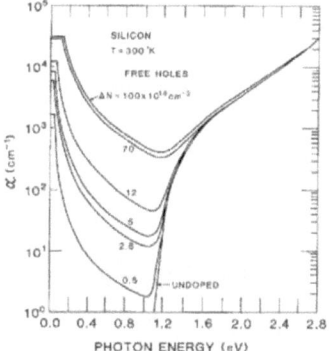

Figure 4.9.: Optical absorption spectrum, various concentrations of free holes [SB87]

the data predicted by the theory are approximately 0.5 of the actual values for holes and 0.25 of the experimental values for electrons [1].

4.5.2. Free carrier refraction

Changes in the index of refraction (Δn) due to free carriers (ΔN) are calculable as follows (see appendix for derivation of the equation):

$$\Delta n = \sqrt{n_0^2 - \frac{(q\lambda)^2}{4(\pi c_0)^2 \epsilon_0} \cdot \frac{\Delta N}{m}}, \qquad (4.5)$$

The free carrier refraction data was extracted from the free carrier absorption data by Soref and Bennett from the graphs shown in figure 4.10 and 4.11. The resulting curves are shown in figure 4.12 and 4.13.

From this set of refraction data versus wavelength, the dependency of the index of refraction on the charge carrier density for free electrons and holes was determined at the two wavelengths: 1300 nm and 1550 nm. The results of 1300 nm are shown in figure 4.14.

---

[1] Note: the equations for free carrier absorption and refraction shown in their publication ([SB87]) differ slightly form the equations shown in this work, which might be a reason by itself that the theory differs from the experimental data. And: investigations of the theory showed that the dependency of the absorption coefficient on the mobility changes the results, depending on which model is chosen for the calculations of the mobility. The model used for the mobility calculations in this work is explained in chapter 12.1.3 and shown in the appendix.

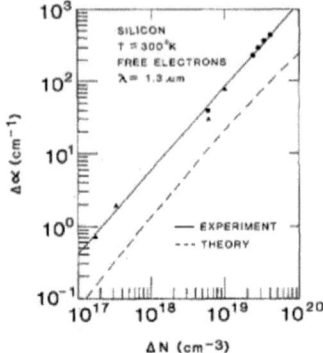

Figure 4.10.: Absorption as a function of free electron concentration [SB87]

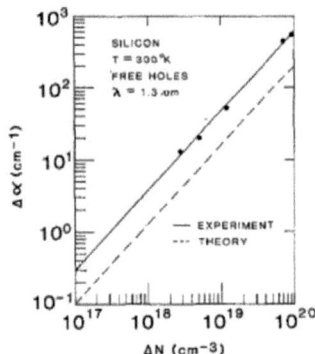

Figure 4.11.: Absorption as a function of free hole concentration [SB87]

Their theory predicts that $\Delta n_h = 0.66 \Delta n_e$ over the entire ΔN-range. The results show $\Delta n_h = 3.3 \Delta n_e$ at 10^{17} cm^{-3}. For electrons, the data is in good agreement with the prediction of the theory, but for holes, the dependency on the free carrier concentration differs from the theory.

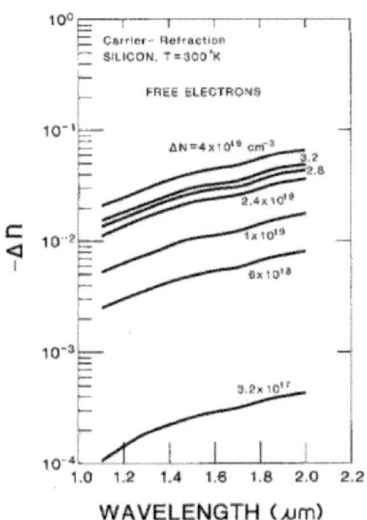

Figure 4.12.: Index of refraction, perturbations produced by various concentrations of electrons [SB87]

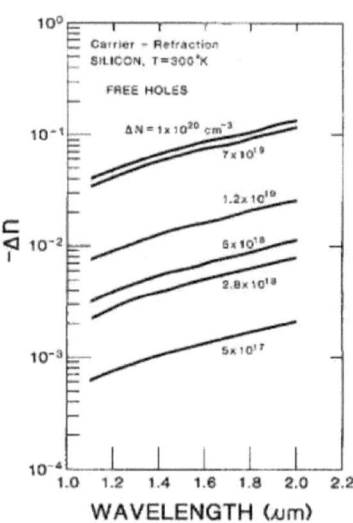

Figure 4.13.: Index of refraction, perturbations produced by various concentrations of holes [SB87]

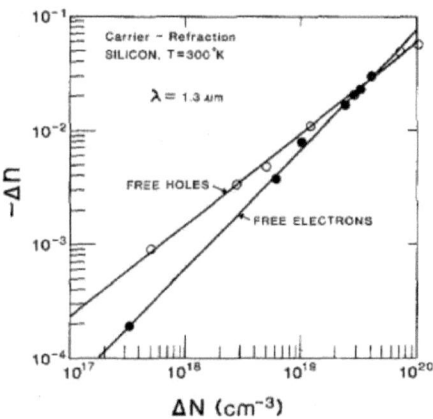

Figure 4.14.: Free carrier refraction as a function of free carrier concentration [SB87]

Part III.
LVP setup and measurement methods

5. LVP setup and components

In part II the relevant effects that influence the reflection from a device were described. This part discusses the setups and measurement methods, which are used to detect these changes in the reflection. Chapter 5 will outline the general LVP setups and the modifications that were introduced for the setup of this work. Further it will discuss topics like the laser wavelengths. In chapter 6, new and conventionally used measurement methods will be summarized and the image acquisition scheme will be explained.

5.1. General overview of LVP setups

The key components of an LVP tool are the laser scanning module (LSM), the data acquisition electronics, the GUI for interaction of the user and the system and of course the laser and the device under test. The commercial systems that are available on the market (e.g. the IDS 2000 from *Credence Systems*, former *Schlumberger*) employ a 1064 nm ML laser and precision timing electronics. To obtain waveforms of digital signal patterns (timing-information) with a ML laser setup, complex sampling and synchronization techniques between a tester and the probing equipment are necessary and the ML laser needs a well-controlled environment (see e.g. [BDM+99]). Usually the acquisition time for one single waveform is about a few minutes. Due to constantly growing demand on system performance, several different setups were described in the literature (e.g. the *Dual Laser Noise Reduction Scheme* or the *Phase Interferometer Detection (PID)*, compare to [BDM+99] and [WL00]), but are not discussed here.

5.2. The LVP setup used for this work

To keep the system simple, for this work, a setup with a cw laser has been used in so-called *Standard Mode* [WL00]. Instead of a tester, a function generator was used to drive the device. For the extraction of the frequency information, a spectrum analyzer (SA) was employed, which reduces the acquisition time drastically (instantaneously to seconds, depending on the data treatment). The schematic diagram of the experimental setup is shown in figure 5.1. Specifications of all the tools used in this work are given in the appendix (14.3). The measurement methods are outlined in chapter 6.

The design used here (Standard Mode, see above) was derived from the PDP mode (polarization difference probing, see [LNB+04]), which uses a common-path interferometer. The optical path of the PDP mode is shown in figure 5.2. A detailed description of the optical path of the PDP mode can be found in [LNB+04]: " [...] two orthogonal,

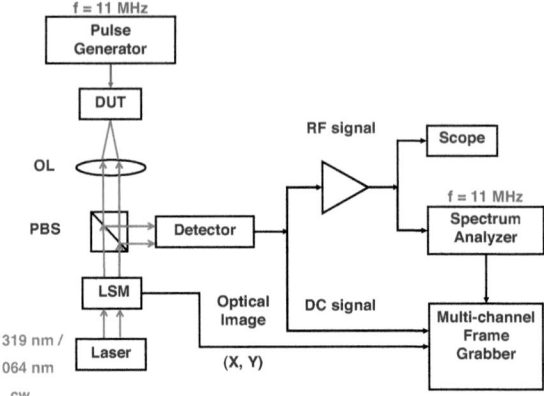

Figure 5.1.: LVP setup for time-domain measurements, voltage sweeping and modulation amplitude mapping, DUT: device under test, OL: objective lens, PBS: polarizing beam splitter, LSM: laser scanning module, cw: continuous wave

linearly polarized laser beams, generated from the same laser, are incident on the same point in the device under test. [...] The beam modulation due to device interaction is unequal for the two beams, because of the polarization dependence of the [...]" reflectance (see section 3.1: σ- and π-case). The two beams interfere after the interaction with the device took place. The incident beam path is shown at the top of figure 5.2. In order to control the reflected beam power, the incident laser power is monitored using the incident pick-off (incident p/o) generated at the first polarizing beam splitter cube (PBS1). "The next PBS cube (PBS2) is oriented to transmit the remaining vertically polarized beam. The beams polarization state is rotated 45° from the vertical by the action of the Faraday rotator (FR); the third PBS cube (PBS3) is oriented to transmit this beam. An equivalent description of the beam at this point is that it is the superposition of a vertically polarized beam and a horizontally polarized beam, both beams equal in amplitude and in phase with each other. The fast and slow axes of the variable-retarder (VR or wave plate) are aligned along these two polarization directions. Thus, after passage through VR, the beam incident on the DUT consists of two spatially coincident, equal-amplitude, orthogonally polarized beams that are phase-shifted (retarded) with respect to each other by a small amount (nominally, $\frac{\pi}{4}$). [...] After retro-reflection by the DUT (see bottom of figure 5.2) the two linearly polarized beams retrace their path to VR. Passing through VR introduces an addition phase-shift between the two return beams (now nominally phase-shifted by $\frac{\pi}{2}$). At PBS3, half of each beam is reflected and half is transmitted. The transmitted halves are deflected out of the beam path via the action of FR and PBS2. The reflected halves interfere, since they are now in the same polarization state, and generate the reflected-A signal. The transmitted halves also interfere for the same reason, generating the reflected-B signal."

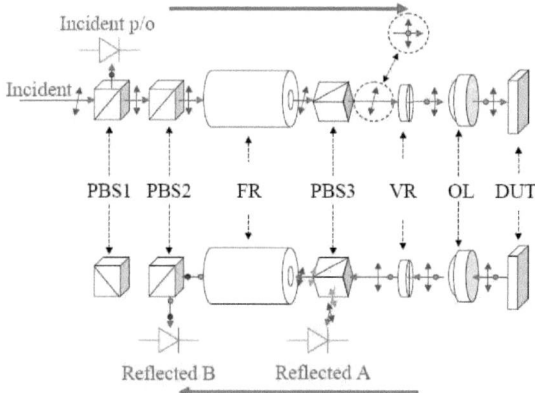

Figure 5.2.: [LNB+04]: "PDP optical path showing the polarization state of the laser beam. Top: Incident beam path, Bottom: Reflected beam path. FR: Faraday Rotator, VR: Variable Retarder, OL: Objective Lens, p/o: Pick-Off. Along the beam path, vertically oriented arrows indicate a vertically polarized beam while dots indicate a horizontally polarized beam. A tilted arrow indicates a beam that is linearly polarized at some angle off the vertical. Spatial separation between a dot and an arrow depicts a phase difference between the two beams. The dotted circles in the incident beam path, between PBS3 and VR, indicate the equivalence between a 45° polarized beam and two in-phase, equal amplitude beams that are polarized vertically and horizontally."

For the standard mode, only the reflected-A is used (in this case the Faraday-Rotator is not necessary). The detected amplitude of the reflected beam thus contains information of the amplitudes from both polarization states plus the interference of both in the form of one single data point (simply called "reflected light" in the following).

5.2.1. Interference of polarization states at the detector

As described above (5.2), the setup only enables amplitude measurements, which means that both components of the electric field (perpendicular and parallel to the plane of incidence) interfere at the detector as described in equation 3.11 in section 3.2.1. As shown in figure 5.2, the reflected-A signal consists of two reflected halves of a beam that interfere, since they are in the same polarization state. The phase shift between the two halves is $\frac{\pi}{2}$, so from equation number 3.11 it can be derived that the interference of the two beams results in a simple addition of the two reflected waves (the difference in path length should be the same for both). The two polarization states, however, may cause different properties for the respective interfering waves due to the effects described with the equations number 3.2 (σ-case) and 3.5 (π-case). In addition, the indices for both polarization states may differ due to electro-refraction as described in 4.4.2, which might alter the amplitudes of the fields, as well.

5.2.2. Laser wavelength (1064 nm versus 1319 nm)

As described in section 4.2, the effects that occur in silicon while a laser beam is focused onto the active areas of the device are depending on the wavelength used. Further it was shown that light with photon energies greater than the band-gap energy is capable of generating electron-hole pairs. This means that lasers with wavelengths lower than 1107 nm (equivalent to 1.12 eV) can influence the signal while the measurement takes place, which should be avoided. For example (see [BDM+99]), this was found in ML laser measurements of pre-charged floating memory cells (timing of the discharge - read cycle - depends on the laser power and the pulse width). However, the same publication described that nA (typical range of induced photo currents) do not affect the waveform timing information noticeably for circuits with non-floating nodes. On the other hand, the resolution plays an important role. The limit of resolution a_{min} (after Abbe) is defined as

$$a_{min} = \frac{\lambda}{2NA}, \tag{5.1}$$

for coherent light and perpendicular incidence (e.g. compare to [KF88] or [Hec87]), where NA is the numerical aperture. This means that to obtain the best resolution it is necessary to either reduce the wavelength or to increase NA. The NA of the lens employed in the system was 0.85. For the measurements of this work, two lasers with different wavelengths were available: a **1319 nm** laser, with a photon energy (0.94 eV) well below the band-gap energy to avoid electron-hole pair generation (**non-invasive**) and a **1064 nm** laser that is capable of generating electron-hole pairs (photon energy 1.17 eV, slightly above the band-gap energy) but decreases the limit of resolution.

6. Measurement methods and image acquisition

A laser is focused to the active areas of the DUT through the thinned backside using a high-magnification lens (100x objective). The beam position is controlled by the LSM. The beam can either be scanned over the sample (to obtain signals for various positions (x,y) and a fixed voltage level, compare to section 6.2.2), or pointed statically to one position (which enables extraction of signal-to-voltage correlations, see section 6.2.2). Light reflected from the DUT is measured by an avalanche photo diode (APD, detector) used in linear mode. (Note that for output voltages higher than 1 V, the APD is not in linear mode any more. This means that the power of the reflected light needs to be kept low for all samples (especially the ones that are highly reflective) and for all areas, where an accurate modulation depth needs to be extracted. This can be achieved by reducing the laser power of the incident beam.)

6.1. Image acquisition

The static or DC part of the signal (R_0) for each pixel is sent to a multi-channel frame grabber card. The frame grabber also uses the (x, y) location information from the LSM to construct an optical image (static part of the reflected light) from the 512x512 pixels - see figure 6.1. To enable fast signal acquisition and image processing, the function generator (set the drive voltages), the SA (read the modulation data) and the multi-channel frame grabber (collect all the data needed to process the images) were controlled by *Labview*.

6.2. LVP signal acquisition

The second part of the signal, $\Delta R(t)$, is the RF component that is modulated by the transistor dynamics (voltage dependence). It is amplified and sent to an SA and an oscilloscope. The modulated part can be inspected in different ways. **A) Time-domain measurement**: the oscilloscope shows the shape and sign of the signal versus time. **B) Frequency-domain measurement**: a spectrum analyzer records the signal amplitude at the frequency of operation (figure 6.2). The oscilloscope and the spectrum analyzer measure only the raw voltage from the APD. In order to understand the physics of LVP it is necessary to convert these measurement data into relative modulation levels $\Delta R(t)/R_0$ [ppm] (see section 6.2.3). If the reflectance due to the device activity increases from the DC value, the LVP signal is said to be "positive", whereas a decrease in the

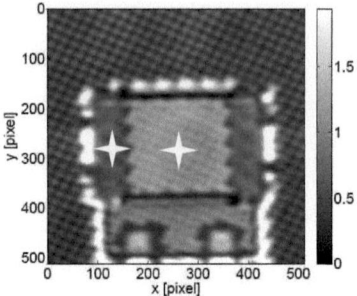

Figure 6.1.: Image of the DUT (static part of the reflected light) - markers show the probe points in the center of the drain (left) and the gate (right)

reflectance results in a "negative" signal. This "sign" information can be achieved from the waveform taken with the oscilloscope, which shows the information for one single position, but can also be extracted with a lock-in amplifier, which is another frequency-domain measurement method, described in section 6.2.2.

6.2.1. Time-domain measurement

Time-domain measurements have been widely used in commercial systems to extract waveforms (signal versus time). This method helps investigating signals of unstable or unknown frequencies, since it is not possible to set the SA to the frequency of operation (compare to section 6.2.2). However, this method is time-consuming, a few minutes per waveform, due to the high latency of the oscilloscope, but it is possible to extract the sign (reflection increasing or decreasing) from the data.

6.2.2. Frequency-domain measurement

The **frequency-domain** measurement is faster and less noisy, due to the narrow resolution bandwidth that can be used. Another advantage of the frequency-domain measurement is that it does not need a trigger signal from the circuit. Only the frequency of the signal needs to be known, then the SA can be set to this specific frequency. This method gives no information about whether the acquired RF signal is increasing or decreasing (sign). In opposition to the time-domain measurements, the measurement methods based on the frequency have not been used in commercial LVP tools yet.

Based on frequency-domain signals, two different types of measurement can be performed:

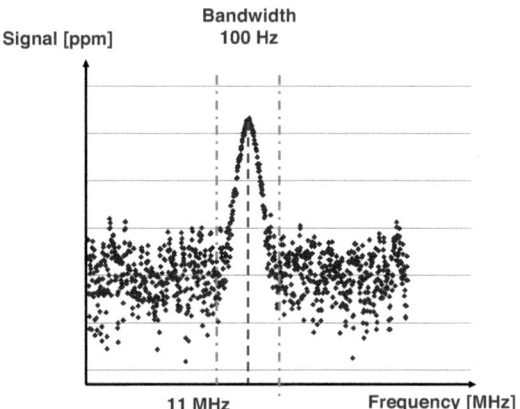

Figure 6.2.: Modulated signal versus frequency (frequency-domain). The amplitude of the RF-signal has been converted into a calibrated modulation depth (parts per million).

Voltage sweeping

One measurement method is the **voltage sweeping** (VS). Here the laser is pointed to a specific position at the device - for example at the drain (see left marker in figure 6.1). One or more terminals of the device are driven by a square wave with a swept amplitude voltage, for example from 0 to 1.2 V with 20 steps. The frequency of the SA is centered at the peak frequency (same as the drive frequency of the device) with zero span. At each drive voltage V_{drive}, the peak signal amplitude measured by the SA, $\Delta R(t)$, is recorded and converted into a ppm value. The result is a diagram that shows the modulation depth of the light versus drive voltage (figure 6.3).

Modulation amplitude mapping

A **modulation amplitude map (MAM)** can be produced by scanning the beam across the device in x- and y-direction. In this case, the drive voltage applied to the device is kept fixed. At each (x, y) pixel location, the ppm value from the SA is recorded by an additional channel on the multi-channel frame grabber card (see figure 5.1). For the acquisition, the scanning speed has to be rather slow to obtain a good signal-to-noise ratio. Usually, for one image of 512x512 pixels, the scan time is around 200 s (the dwell time of the beam at each pixel thus is about 1 ms). In this way, a two-dimensional image of the signal modulation depth over the scanned area is recorded (see figure 6.4

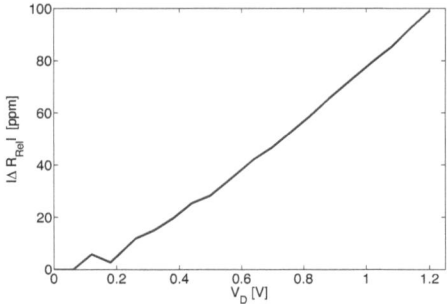

Figure 6.3.: Voltage sweeping: the laser is pointed to one specific position at the DUT, here the drain, the amplitude of the drain voltage V_D is swept and the LVP signal ($|\Delta R_{Rel}|$ in ppm) is recorded

b)). Simultaneously the static part of the reflected light is recorded to provide an aligned optical image (6.4 a)). A similar technique has been reported in the literature, using sample scanning instead of beam scanning [LFB+05].

Modulation sign mapping

For the acquisition of a **modulation sign map (MSM)**, the setup shown in figure 5.1 needs to be modified: instead of the SA a lock-in amplifier is used to evaluate the signal. The trigger signal for the lock-in amplifier is taken from the same source as the drive signal for the DUT (coming from the function generator, having the same frequency as the drive frequency). The modified setup is shown in figure 6.5. Similarly to the modulation amplitude maps, the MSM is performed by scanning the laser beam across the structure while the drive voltage of the device is kept fixed and the values from the lock-in amplifier are recorded by the multi-channel frame grabber. A MSM shows for each position (x, y), whether the reflection of the device is decreasing (black in the image) or increasing (white in the image) with the device voltage (sign). A showcase MSM of a device can be found in figure 6.4, too (c)).

6.2.3. ppm-calculation

As mentioned above, the oscilloscope and the SA measure only the raw voltage from the APD. To extract reflection information that can be correlated to the applied voltage, relative modulation levels (R_{Rel}) need to be achieved. Since both, the static and the modulated part of the reflected light P_{ref}, are depending on the incident beam power P_{inc} (the higher the incident beam power, the higher the reflected power), the relative modulation data is independent of the incident beam power:

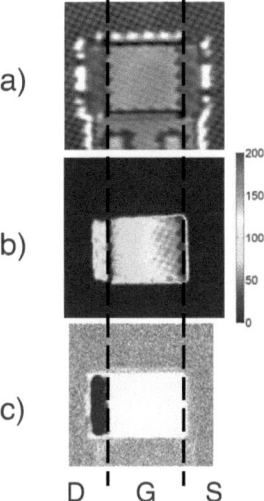

Figure 6.4.: a) Image, b) MAM with ppm values and c) MSM (white: reflection increases, black: reflection decreases) for the same structure (dotted line is showing the areas of drain (D), gate (G) and source (S))

$$R(t) = \frac{P_{ref}(t)}{P_{inc}} = R_0 + \Delta R(t), \quad (6.1)$$

$$\frac{R(t)}{R_0} = \frac{R_0 + \Delta R(t)}{R_0} = 1 + \frac{\Delta R(t)}{R_0}. \quad (6.2)$$

All the modulation maps and voltage sweeps shown in part V are converted to this relative modulation depth $|\Delta R_{Rel}| = \left|\frac{\Delta R(t)}{R_0}\right|$ and the values are in parts per million [ppm].

Data treatment

One problem that arises from converting the modulation data - as described above - is a devision by zero, caused by the reflected light values. Similarly, artificially strong modulation peaks would be detected, when the reflected light intensity is close to zero in areas, where the image (R_0) appears dark. Detailed investigations of post-processing

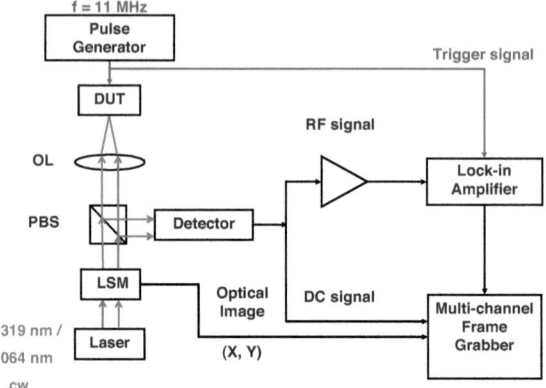

Figure 6.5.: LVP setup for modulation sign mapping, DUT: device under test, OL: objective lens, PBS: polarizing beam splitter, LSM: laser scanning module, cw: continuous wave

of the modulation data were performed with *Matlab*, concluding to calculate the ppm-signal only, if the (linear) modulation raw data $\Delta R(t)$ is above a threshold (20%) of the overall maximum modulation data, otherwise the ppm-value was set to 0. This way, the ppm-values are only calculated in areas, where the modulation is strong due to the device activity. Figure 6.6 shows two modulation maps in comparison - a modulation map with and without post-processing of the modulation data as explained above.

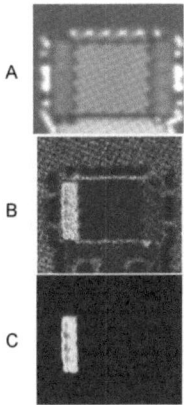

Figure 6.6.: Comparison modulation maps: calculated ppm-values (B) (artificially high in areas, where the reflected light is close to zero (A)) versus same data with post-processing and smoothing of the data (C)

Part IV.

Devices

7. Overview of the devices used

All devices investigated for this work were test chips from Infineon Technologies AG. The devices - MOSFETs - were chosen for various reasons. Mainly, the scope of this work is to understand better the signal origin of LVP signals, to forecast the expected signal levels for future technologies and scaling and to investigate the usability of different laser wavelengths (see section 5.2.2).

All of the transistor terminals are accessible separately, which allows investigating drain and gate signals independently from each other (measurements of the FETs in various operation modi will be described in part V). To get to a conclusion, in which area of the device (particularly in gate and drain areas) the signal is generated, **large** test structures, in **120 nm** process technology, such as FETs with gate lengths (L) and widths (W) of 10 μm, have been chosen. In the following chapters, these FETs will be abbreviated as: **NF-L-120** (**NFET**) and **PF-L-120** (**PFET**). The abbreviations of all FETs that were used follow the same pattern.

Sub-micron ($\frac{W}{L} = \frac{0.16 \mu m}{0.12 \mu m}$) devices, also in 120 nm technology, - **NF-S-120**, **PF-S-120** - and devices in 65 nm process technology - **NF-L-65** and **PF-L-65** - have been measured, to investigate the signal behavior with the scaling and to determine, whether there are differences between two process technologies.

The two process technologies (120 nm and 65 nm) differ in several parameters. The thickness of all the layers are decreased; e.g. the gate oxide and the diffusion depths. The doping concentration is increased and the doping profiles are changed. Even the materials used - such as the silicide material characteristics - are varied and the aspect ratio is changed. In terms of the layout, there is only a small difference between the two process technologies that is the well, which is made separately for each FET in 65 nm technology, where it expands across the entire macro for the 120 nm technology (see section 7.2). With all these modifications, the electrical data changes as well, which will in turn influence the measurements, shown in part V. All devices were measured with the two laser wavelengths available (1319 nm and 1064 nm) as will be described in chapter V.

7.1. Electrical characteristics of the devices

The nominal device voltage of all N- and P-FETs is $V_{NOM} = \pm 1.2$ V. Output- and transfer characteristics have been measured for each device. The threshold voltages V_{thr} and drain saturation currents $I_{D,sat}$ (or I_{ON}) can be found in table 7.1. For the modeling of the NF-L-120 and the PF-L-120, the built-in potential V_{bi} needs to be known as well. It was measured and determined to be ±0.81 V (positive for the PFET and negative for the NFET).

7.2. Layouts of the devices

Table number 7.2 shows the layout of the FETs used: N/PF-L-120, N/PF-S-120 and N/PF-L-65. The layouts of the NFETs and the PFETs are the same in both technologies - there are wells for both, only with opposite doping type -, so only one layout of the two FETs is shown. The colors in all layouts were chosen to be the same, for better understanding: red: poly-Si, blue: metal 1, white: diffusion areas, yellow: contacts (can not be resolved in the layout of the NF-L-65), orange: well. The dimensions are labeled in each layout. The well of the devices in 120 nm technology is larger in size than the well of the 65 nm technology and thus can not be seen with the chosen zoom factor. The FETs are surrounded by STI (shallow trench isolation), which does not appear in the layout. There are also STI and metal fills (dummy fills), but like the STI they are not shown in the layout. The images of the devices are shown and discussed in detail in chapter 8. In that chapter, the location of the STI and the fill shapes will be explained with the help of the images. In addition, a schematic of a cross-section of a device will be presented.

7.3. Technology parameters

The exact technology parameters, like the doping profiles, cross-sections of devices that reveal detailed information about the dimensions etc., are confidential data of Infineon Technologies AG and thus can not be shown. However, the following paragraph contains approximations of the thicknesses of the layers (substrate, well, diffusion, silicide, nitride, gate oxide and poly-Si, see table 7.3), the doping concentrations (substrate, well, diffusion and poly-Si) and a schematic of a FET (see figure 7.1) of the 120 nm process technology as an example. Apart from the doping type, the parameters of the NFET

Table 7.1.: Electrical characteristics of the devices used: threshold voltage V_{thr} and drain saturation current $I_{D,sat}$

Device	V_{thr} [V]	$I_{D,sat}$ [μA]
NF-L-120	+0.22	+150
PF-L-120	-0.22	-31
NF-S-120	+0.27	+75
PF-S-120	-0.24	-30
NF-L-65	+0.20	+154
PF-L-65	-0.19	-29

and the PFET are very similar, so the data apply for both unified. The approximations of the thicknesses of the layers and the carrier concentrations will be used in chapter VI for the simulations of the reflectance from a FET.

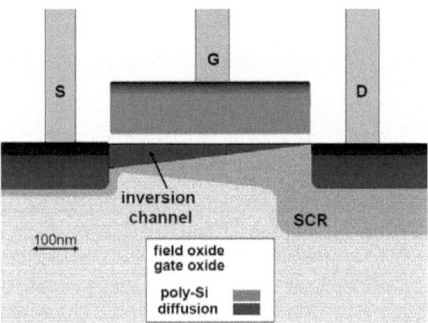

Figure 7.1.: Schematic diagram: FET with contacts to the active level (S: source, G: gate and D: drain), diffusion, poly-Si and oxides as labeled in the figure

For the approximation of the doping levels in the FETs - substrate, well, diffusion and poly-Si - a data set of the electrical data simulation with DESSIS ISE TCAD has been used. The extracted average of the substrate doping concentration is $N_{sub} = 5 \cdot 10^{15} cm^{-3}$, the well doping concentration about $N_{well} = 2 \cdot 10^{17} cm^{-3}$ and the diffusion doping concentration $N_{diff} = 2 \cdot 10^{20} cm^{-3}$.

7.4. Device modi

For the device modeling and the simulations of the reflectance from a FET (see part VI), approximations of the charge carrier densities (as above for the doping levels) and the according thicknesses of the layers in the FETs are necessary. Especially for the modeling of the active signal contribution (compare to section 10) the properties of the inversion channel and the SCR need to be calculable. In the following sections, these parameters will be discussed for the NF-L-120 and the PF-L-120 (devices used for the modeling) and an explanation will be given, how the calculations will be included in the simulations.

7.4.1. Reverse biased diode

When only the drain of a FET is pulsed with the nominal device voltage (positive for an NFET and negative for a PFET) and gate, source and well are grounded, the diffusion-to-well junction of the FET can be understood as a diode in reverse bias. For an abrupt pn junction and constant doping profiles, the thickness of the belonging SCR can be calculated as follows ([SN07]):

$$t_{SCR} = \sqrt{\frac{2\epsilon_0 \epsilon_{Si}}{q} \left(\frac{1}{N_{diff}} + \frac{1}{N_{well}} \right) (V_{bi} - V_j)}. \quad (7.1)$$

Wherein t_{SCR} is the thickness of the SCR, N_{diff} and N_{well} are the diffusion and well doping concentrations, in m^{-3} so the thickness of the SCR is calculated in m, V_{bi} is the built-in potential (see below) and V_j is the voltage applied to the junction, e.g. 0-1.2 V for the NFETs, called V_D in the simulation results, because it is applied to the drain.

The built-in potential is defined as:

$$V_{bi} = \pm V_t \ln \left(\frac{N_{well} N_{diff}}{n_i^2} \right). \quad (7.2)$$

Here $V_t = \frac{kT}{q}$ is the thermal voltage, n_i^2 is the intrinsic carrier density squared and is equal to $1.71 \cdot 10^{20}$ cm^{-6}.

The built-in potential was also *measured* for the NF-L-120 and the PF-L-120, see section 7.1. To fit this measured value to the calculable V_{bi} with the extracted well and diffusion doping concentration from above, a fitting parameter x will be introduced in the equation, as well:

$$V_{bi} = \pm x V_t \ln \left(\frac{N_{well} N_{diff}}{n_i^2} \right). \quad (7.3)$$

With this equation the thickness of the SCR can be evaluated for various doping concentrations as will be discussed in chapter 12.

7.4.2. Varactor in inversion

When only the gate of a FET is pulsed and drain, source and well are grounded, the material system poly-Si gate (in former times **metal, M**) / gate oxide (**oxide, O**) / **semiconductor (S)** in the gate area can be understood as a **MOS**-varactor (voltage dependent capacitance) in inversion.

In this work, the effects of substrate bias (potential difference between source and bulk; V_{SB}) are not discussed. For a detailed description of the according threshold voltage shift and the influence on the SCR thickness see [SN07].

For voltages below the threshold voltage V_{thr} and a constant doping profile, the SCR thickness can be estimated from equation number 7.1, substituting the drain voltage with the gate voltage and only taking "half" of the pn junction (only well doping concentration) into account. The effects of fixed oxide charges and the work function difference between the gate material (poly-Si) and the semiconductor of the substrate are included in the so-called flat-band voltage V_{FB}, which is required by the gate to produce the flat-band condition (compare to [SN07]). Accordingly the charges in the gate stack cause an initial SCR, even for a gate voltage V_G of zero volts. This initial SCR thickness $t_{SCR,initial}$ was estimated to be 50 nm and has to be taken into account for the calculation of the overall thickness, as well. This $t_{SCR,initial}$ is an approximate value: a rough estimation based on process technology experts knowledge - around half the thickness of a source / drain diffusion. These 50 nm correlate with a flat-band voltage of 0.38 V. In principle, the initial SCR thickness at zero volts is, as equation number 7.1 shows, also depending on the well doping concentration. For simplicity reasons, only the general influence of the well doping concentration on the **overall** SCR thickness will be evaluated in the simulations of this work in chapter 12 (the influence on the **initial** SCR thickness will be neglected). Consequently the following equation will be used to calculate the thickness of the in the varactor:

$$t_{SCR} = t_{SCR,initial} + \sqrt{\frac{2\epsilon_0\epsilon_{Si}}{q}\left(\frac{1}{N_{well}}\right)|V_G|}. \tag{7.4}$$

After the threshold voltage is reached, the SCR thickness stays constant at this level and does not increase any further.

For the approximation of the inversion channel charge carrier density N_{chan}, the data set of the electrical data simulation with DESSIS ISE TCAD has been used. For the calculations in this work, it will be assumed that for gate voltages below the threshold voltage the inversion channel charge carrier density is zero [1]. In this work, for $V_G \geq V_{thr}$, the carrier density in the channel will be calculated with the following equation (polynomial fit to the data of the electrical simulation, result in cm^{-3}):

$$N_{chan} = -8.7 \cdot 10^{18} |V_G|^4 - 6.8 \cdot 10^{18} |V_G|^3 + 1.3 \cdot 10^{20} |V_G|^2 - 2.2 \cdot 10^{19} |V_G|^1 + 3.3 \cdot 10^{17}. \tag{7.5}$$

For the variations of the inversion channel charge carrier density in chapter 12 this equation is simply multiplied by a certain factor.

The threshold voltage is calculable by the following equation (see [SN07], eq. 25):

$$V_{thr} = V_{FB} + 2\Phi_B + \frac{\sqrt{4\epsilon_{Si}\epsilon_0 q N_{well}\Phi_B}}{C_{ox}}. \tag{7.6}$$

Here, V_{FB} is the flat-band voltage, $\Phi_B = V_t ln\frac{N_{well}}{n_i}$ is the bulk potential and $C_{ox} = \frac{\epsilon_0\epsilon_{SiO_2}}{d}$ is the oxide capacitance per unit area (with $\epsilon_{SiO_2} = 3.9$ and d, the thickness of

---

[1]The threshold voltage is defined as the voltage level, at which the minority carrier density (free carriers) at the surface of the semiconductor is equal to the majority carrier density in the unperturbed volume. For voltages higher than the threshold voltage, the inversion charge carrier density is a strong function of the applied gate voltage.

the gate oxide, see table 7.3). This equation shows that the threshold voltage - as the built-in potential - depends on the well doping concentration. Chapter 12 will investigate the influence of variations in the well doping concentration on the simulations, for which the threshold voltage needs to be estimated. For the simulations, the threshold voltage was *not* calculated directly from equation 7.6 in order to match the measured value. The first order approximation (only the trend of the threshold voltage needs to be evaluated, as the simulations will show) of the doping influence on the threshold voltage proceeded as follows. The measured threshold voltage of the devices used for the modeling (± 0.22 V for NF-L-120 and PF-L-120, see table 7.1) correlates with the extracted $2 \cdot 10^{17}$ cm^{-3} well doping concentration. The calculated threshold voltage in the simulations (V_{thr}) should match the measured value ($V_{thr,initial}$) for the extracted well doping concentration ($V_{thr} = V_{thr,initial} \cdot y$, with $y = 1$ for a well doping concentration of $2 \cdot 10^{17}$ cm^{-3}).

Equation 7.6 was used to estimate the influence of the well doping concentration on the threshold voltage. The equation consists of three summands: the flat-band voltage, the potential difference in the space charge region at strong inversion ($2\Phi_B$), due to the depletion charge, and the potential difference across the oxide ($\frac{\sqrt{4\epsilon_{Si}\epsilon_0 q N_{well} \Phi_B}}{C_{ox}}$), caused by the surface charge. To give examples, the three summands will be evaluated for three well doping concentrations: $1 \cdot 10^{16}$ / $1 \cdot 10^{17}$ / $1 \cdot 10^{18}$ cm^{-3}.

The flat-band voltage can be understood as the work function difference (neglecting the oxide charge for this examination). The influence of the well doping concentration on the work function difference is described by [SN07] (Work-function difference versus doping, for gate electrodes of degenerate poly-Si on p- and n-Si). The extracted values for well doping concentrations of $1 \cdot 10^{16}$ / $1 \cdot 10^{17}$ / $1 \cdot 10^{18}$ cm^{-3} are -0.9 / -0.98 / -1.04 V for n-poly-Si (gate) and p-Si (substrate) and 0.8 / 0.85 / 0.9 V for p-poly-Si and n-Si. The bulk potential can be calculated directly from the equation shown above. The resulting values are 0.35 / 0.41 / 0.47 V for $1 \cdot 10^{16}$ / $1 \cdot 10^{17}$ / $1 \cdot 10^{18}$ cm^{-3} well doping concentration respectively. The evaluation of the third summand, representing the inversion charge, leads to 42 / 145 / 489 mV for the three well doping concentrations. This shows that the **influence** on the threshold voltage of the third summand is the strongest. Simplifying the third summand further leads to a proportionality of $\sqrt{N_{well}\Phi_B}$, so the influence of the well doping concentration on the threshold voltage can be roughly estimated from the initial values for various doping concentrations from equation 7.7:

$$V_{thr} = V_{thr,initial} \frac{\sqrt{N_{well}\Phi_B}}{\sqrt{N_{well,initial}\Phi_{B,initial}}}. \tag{7.7}$$

Where $V_{thr,initial} = 0.22$ V, $N_{well,initial} = 2 \cdot 10^{17} cm^{-3}$, $\Phi_{B,initial} = V_t ln\frac{N_{well,initial}}{n_i}$ and the according well doping concentrations N_{well} for which V_{thr} should be estimated. Note that this estimation predicts very strong variations of the threshold voltage with the well doping concentration, which does not represent the real influence too well, but it will show the trend of the dependency in the simulations clearly.

7.4.3. FET, gate and drain simultaneously pulsed

When both pins - gate and drain - of a FET are pulsed simultaneously and source and well are grounded, the FET is switching between off-state and on-state, in which the drain current is flowing from source to drain. In this device mode, in the gate area, the SCR thickness decreases from drain to source and the inversion channel thickness decreases from source to drain.

Table 7.2.: Layouts of the devices used (red: poly-Si, blue: metal 1, white: diffusion areas, yellow: contacts, orange: well)

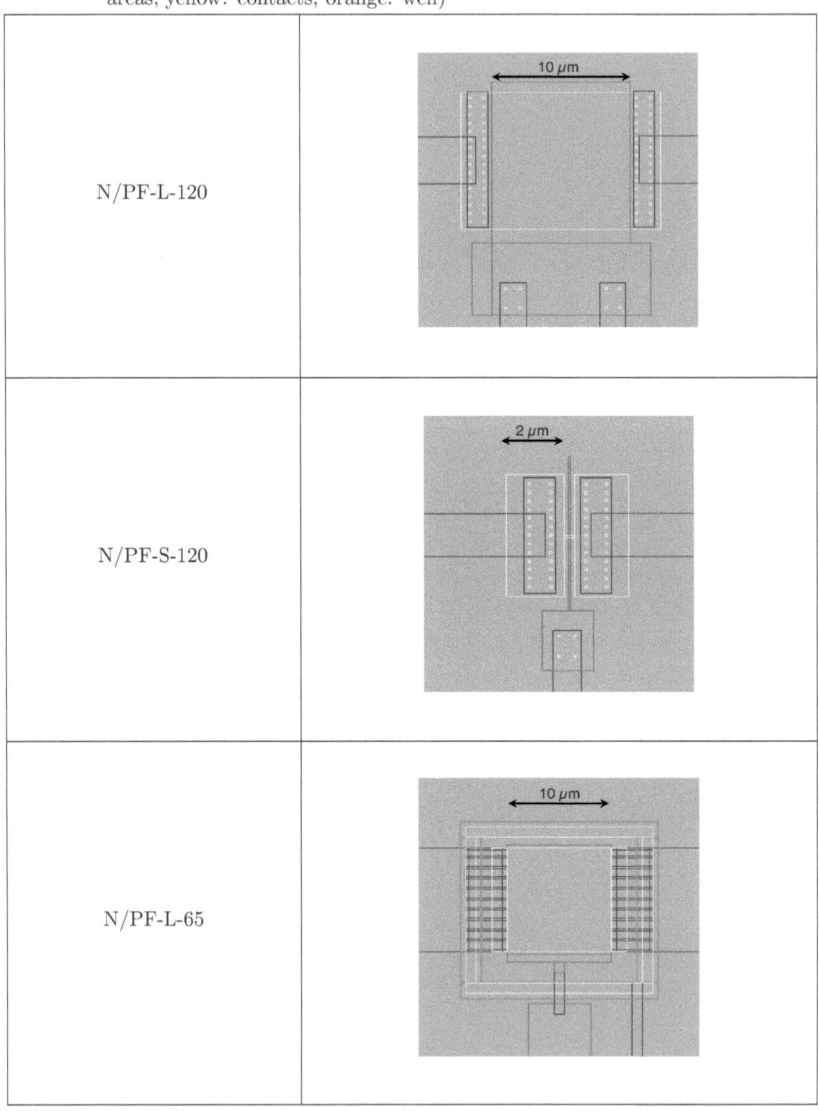

N/PF-L-120	
N/PF-S-120	
N/PF-L-65	

Table 7.3.: Thicknesses of layers occurring in a FET of the 120 nm process technology, rough approximations

Layer	Thickness
substrate	100 μm
well	2 μm
diffusion	100 nm
silicide	25 nm
nitride	25 nm
gate oxide	3 nm
poly-Si	100 nm

Part V.

Measurements

In this part, the results of the LVP measurements of all devices will be described and compared. The differences of the signals between the types of FETs (NFET versus PFET), the laser wavelengths used for the measurements (1319 nm versus 1064 nm), the process technologies (120 nm versus 65 nm process technology) and sizes (large structures with 10 μm gate length and width versus nominal gate length and width) will be discussed.

First, the static part of the reflected light that was recorded during each measurement will be evaluated (chapter 8). This is an infra-red optical micrograph of the structure through the backside of the chip (image, compare to 6.1). Chapter 9 will describe the measurements that were performed, while the devices were operating (modulated part of the reflected light).

8. Detailed investigation of the images

The image of the NF-L-120 taken with the 1064 nm laser is shown in figure 8.1 (left). For a better understanding, the layout is shown, too (right). Note that the images of the PFETs and the NFETs are very much alike (see images in chapter 9) and hence only the images of the NFETs are shown here. In the case of the 10 μm structure, the source, the drain and the gate area can be distinguished from each other - see annotations in the figure. It is possible to see the active areas of the device, because the well is almost completely transparent. The laser beam is partially reflected at each interface of the device.

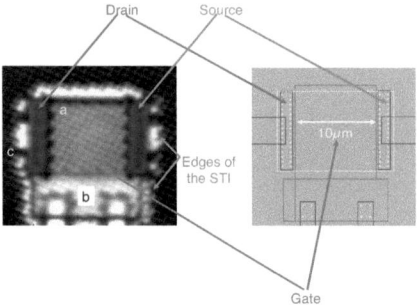

Figure 8.1.: Image of NF-L-120 (PF-L-120 much alike), 1064 nm laser wavelength, (left) and layout (right, compare to table 7.2) showing source, gate and drain. a) poly-Si gate and b) STI are partially transparent, because other structures can be seen through, c) metal in the background (out of focus) still contributes to the signal / image

The FET is surrounded by STI, which shows up the brightest in this image. STI fill shapes surround the STI (this was proven by the preparation of one device with the FIB, not shown, confidential data). The diffusion-to-contact interface is presumed to dominate the reflected signal and provide the contrast in the drain areas. In the drain areas, the metal interconnect layer (M1, blue in the layout) in the background of the FET is almost out of focus, but does at least partially contribute to the signal / image: a weak pattern is detectable in the background of the diffusions and the reflection in the STI area decreases a bit, when the edges of the metal lines are in the background (see c) and b) in the image). The gate area (labeled a)) is brighter than the diffusions. Especially in this image (1064 nm laser), a pattern can be found in the center of the gate. A preparation of an already thinned device with the FIB has been performed: etching further and further, the well, the diffusions, the gate oxide, the poly-Si gate and all the following metalization layers (M1-M5) were removed. The images taken during the preparation (not shown, confidential data) revealed that the metal fill shapes fit the pattern found in the center of the gate area in the optical images in size and shape. So the only conclusion is that the gate stack (poly-Si plus silicide) is transparent to a degree that a reflection pattern of structures beyond the poly-Si gate is transmitted. A replication of a cross-section, also done with the FIB, shows the situation (see figure number 8.2).

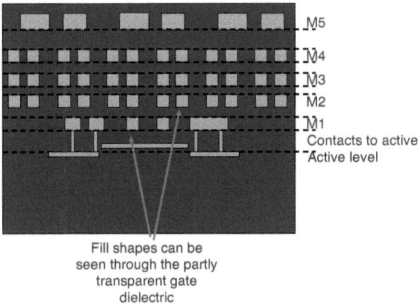

Figure 8.2.: Replication (schematic) of the cross-section that was done with the FIB showing the metal layers (M1 - M5) above the active area of a FET. Fill shapes can be seen through the partly transparent gate dielectric.

The NF-L-120 image taken with the 1319 nm laser (compare figure 8.3 a) and b)) clearly differs from the 1064 nm measurement, not only in the resolution but also in the amplitude of the signal. In some places the 1064 nm laser image is brighter (gate area, STI), in other areas it is darker (source and drain diffusions, STI fill shapes surrounding the FET). The variations in the level of the reflected light arise from interference effects caused by the interaction of the laser light with the structure. So for example even though the STI is highly reflective, which is the case for both wavelengths, in the 1319 nm image it appears dark in some areas surrounding the FET, where it is bright in the 1064 nm image (compare to chapter 3.2).

In comparison to the devices of the 120 nm process technology, the images of the 65 nm process technology again are distinct. As an example, figure 8.3 c) also contains an image, which was performed with the 1319 nm laser. Figure number 8.4 explains the structure, comparing the image (left) with the layout (right). The two images, which

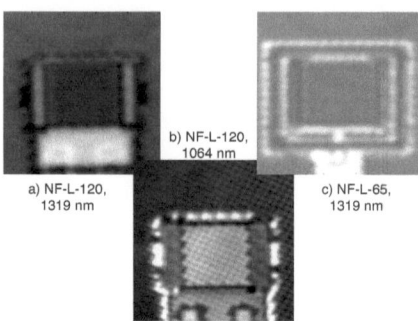

Figure 8.3.: Images: a) NF-L-120 measured with 1319 nm b) NF-L-120 measured with 1064 nm c) NF-L-65 measured with 1319 nm

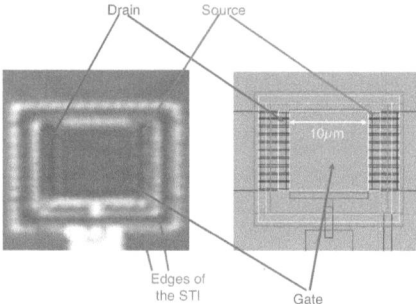

Figure 8.4.: Image, 1319 nm laser wavelength, (left) and layout (right) of the NF-L-65 showing source, gate and drain.

were performed with the 1319 nm laser, - see figure 8.3 a) NF-L-120 and c) NF-L-65 - are the complement of each other (e.g., in the image of the NF-L-65 the STI is bright instead of dark, the diffusions are darker than the gate area etc.). The images of the NF-L-65 / 1319 nm and the NF-L-120 / 1064 nm are more alike - see figure 8.3. Since every single FET of the 65 nm technology has a separate well (see figure 8.4), where the 120 nm process technology had one well for all the FETs (compare to section 7.2, the well for the 120 nm technology is larger in size), the images already differ due to this fact alone: gate and diffusions are surrounded by STI, then follows the well diffusion, which is again surrounded by STI. The additional two rings surrounding gate and diffusions in the image of the NF-L-65 can be traced back to these variations in the layout of the two process technologies.

Summary Even for the same process technology, the images of the two laser wavelengths differ in amplitude of the static part of the reflected light, because the interference effects are distinct. A comparison of one image with the cross-section through the device shows that layers, which seem to be out of focus, still contribute to the signal. When there are variations in the static part of the reflected light for the different devices and wavelengths used, it is likely that the modulated part of the reflected light will differ, too.

9. Detailed investigation of modulation amplitude maps (MAM), modulation sign maps (MSM) and voltage sweeps (VS)

In the following paragraphs, various modes of operation of the FETs will be discussed separately - for the two types of FETs (NFET and PFET) and the two wavelengths (1319 nm and 1064 nm). First, only the drain related LVP signals will be shown (drain diode in reverse bias; only the drain is pulsed, see section 9.1), followed by the gate related signals (varactor in inversion; only the gate is pulsed, section 9.2). After that, the results of the measurements with both pins pulsed simultaneously (FET in switching condition, section 9.3) will be discussed and compared to the previous two modi of device operation. For the voltage sweeps the laser was pointed to the center of the areas being probed: e.g. to the center of the drain (for the diode in reverse bias). When both pins were pulsed, two voltage sweeps were performed: one for each area - gate and drain. The terminals of the device were driven by a square wave with a swept amplitude voltage from 0 to ± 1.2 V with 20 steps. For the MAMs and the MSMs, the drive voltage applied to the device is kept fixed at the level of V_{NOM}.

In the evaluations of the results, the signal *levels* will be discussed. As can be found in section 6.2.2, it is possible to extract signal levels from the MAMs or the VSs. In the paragraphs below, the term *level* will be used to describe the values of the signals at the nominal device voltage (signal strengths at lower voltages will be evaluated in the sections "signal-to-voltage correlations"). The measurements will show slight variations in the signal levels between the VSs at V_{NOM} and the MAMs, which were performed for the same voltage. This is caused by the probe placement - see MAMs: slight variations in the signal levels can be found. Only, when there is a bigger discrepancy between the two levels (as e.g. in the measurements of the devices with the nominal gate dimensions), the reason will be outlined.

In the sub-sections "Different device sizes" of each section, the PF-L-120 will be compared to the PF-S-120 - the device with the nominal gate dimensions. In those sections, only the **PFETs** will be used as a showcase of the results, because the measurements of the **NF-S-120** and the **PF-S-120** are very similar. The only difference in the results of both sub-micron devices that was found has its origin in the sign of the drain signal. This particular difference will be discussed in section 9.4.1.

9.1. Reverse biased drain diode ($V_G = V_S = V_W = GND$; V_D pulsed)

For the measurements in this paragraph, only the drains of the FETs were pulsed. Gate, source and well were grounded. The PFET drains have been driven with a negative voltage, whereas the voltage for the NFET drains was positive. So the drain-to-well junctions of the FETs can be understood as diodes in reverse bias (compare to section 7.4.1). All measurements were performed as described above. For the voltage sweeps, the laser was pointed to the center of the drain diffusions.

9.1.1. 120 nm process technology

Table 9.1 and 9.2 show the results of the measurements for NF-L-120 and PF-L-120 with the according drains driven as reverse biased diodes. Both, NFET and PFET, have been measured with 1319 nm and 1064 nm laser wavelength.

The tables contain the images a), the ppm MAMs performed at V_{NOM} (all maps set to the same ppm-scale - easier comparison) b), the respective MSMs c) and the results of the VSs of all these measurements including the maximum signal level, measured at the nominal device voltage in the VSs d).

b) MAMs (shape of the signals) The MAMs of these drain signals are very much alike: the complete drain diffusion area is contributing to the signal, because - while the drain is driven with a pulse - the width of the SCR of the diffusion-to-well junction is modulated (only active signal source, compare to chapter 10) and hence producing the signal. In the 1319 nm MAMs, there are areas directly surrounding the drain diffusions, which show no signal and, further away from the drain, a ring shaped signal seems to reappear. The areas within the drain diffusions, where the signals are stronger, are caused by the contrast of the reflected light (the images are darker in those places and therefore the ppm-values are stronger, see section 6.2.3, ppm calculation).

c) MSMs (shape of the signals) In the MSMs, the signals surrounding the drains are continuous - there is no gap between the two signals. The maps show that there are sign flips between the signals inside the drain diffusion areas and the "outside" ring shaped signal. For the MAM, only the amplitude of the signal is evaluated, so the signal value can be zero, when it is eliminated by two competing signals of opposite sign that interfere. The areas in the MAMs, where the signal is zero, are caused by this effect (compare to MSMs). The sign of the signals from the NFET diffusions (inner signals) is changing with the laser wavelength (positive for 1319 nm, negative for 1064 nm), whereas the sign is the same in both cases for the PFET (both times negative).

d) VSs (signal-to-voltage correlation) The signal-to-voltage correlations of all drain signals are linear. The signals were measurable down to very low voltages: ± 0.12 V in this case.

Table 9.1.: Reverse biased drain diode: Images, MAMs and MSMs
Modulated: drain ($V_D = \pm 1.2V$ pulsed)
Constant: -
Grounded: gate, source, well ($V_G = V_S = V_W = GND$)

Wavelength / Device	1319 nm NF-L-120	1064 nm NF-L-120	1319 nm PF-L-120	1064 nm PF-L-120
a) Image				
b) MAM				
c) MSM				

Table 9.2.: Reverse biased drain diode: VSs
  Modulated: drain (V_D pulsed)
  Constant: -
  Grounded: gate, source, well ($V_G = V_S = V_W = GND$)

Wavelength / Device	1319 nm NF-L-120	1064 nm NF-L-120	1319 nm PF-L-120	1064 nm PF-L-120
d) VS				
d) VS-level at V_{NOM}	D: 100 ppm G: -	D: 128 ppm G: -	D: 64 ppm G: -	D: 145 ppm G: -

b) + d) Signal levels (MAMs and VSs) The signal levels obtained with the 1064 nm laser are higher than those measured with the 1319 nm laser. The differences between the signal levels is bigger for the PFET: here the result measured with the 1064 nm laser is more than a factor of two higher (145 ppm) than with the 1319 nm laser (64 ppm). The signal levels for the NFET do not differ that much: 100 ppm for 1319 nm and 128 ppm for 1064 nm.

Summary Since the only active signal source of the FETs, while the drains are driven as reverse biased diodes, is the SCR of the diffusion-to-well junction, the signals in the MAMs are similar: they are almost evenly distributed across the drain. The signal-to-voltage correlation is linear in all cases. However, the signs of the signals differ for the two wavelengths and the types of FETs. The signal contribution of the drain signals will be discussed in detail in section 9.4.1.

9.1.2. 65 nm process technology

As explained in chapter 7, there are various modifications from one process technology to the next. To compare the results of the two process technologies - 120 nm and 65 nm - here an NFET and a PFET with 10 μm gate length and width were chosen again (compare to the measurements in table 9.1 and 9.2). The tables 9.3 and 9.4 show the results of the measurements for NF-L-65 and PF-L-65 with the according drains driven as reverse biased diodes. Both, NFET and PFET, have been measured with 1319 nm and 1064 nm laser wavelength.

The tables contain the images a), the MAMs performed at V_{NOM} (all maps set to the same ppm-scale - easier comparison) b), the respective MSMs c) and the results of the VSs of all these measurements including the maximum signal level measured at the nominal device voltage extracted from the VSs d).

b) MAMs (shape of the signals) Except the PF-L-65, measured with 1319 nm laser wavelength, the results are similar to the measurements of the NF-L-120 and the PF-L-120 in table 9.1: the entire drain area produces signals. The signals outside the drain diffusion areas of the PF-L-65, measured with 1319 nm laser wavelength - the ring shaped structures (well) - are caused by the contrast of the reflected light (as explained above, in section 9.1.1 and 6.2.3, the image is very dark in those areas and thus the ppm-value is artificially high). The PFET, measured with 1319 nm laser wavelength shows no signal in the center of the diffusion, which is not caused by the contrast of the reflected light, but by the sign of the signals (explanation see paragraph c)).

c) MSMs (shape of the signals) Except the measurement with the 1319 nm laser wavelength of the PF-L-65, all the signs inside the diffusion areas are positive. The NF-L-65, measured with 1319 nm and the PF-L65, measured with 1064 nm, partly show negative signals surrounding the drain diffusions. In the MAM of the PF-L-65 b), measured with 1319 nm laser wavelength, there is almost no signal in some places of the

Table 9.3.: Reverse biased drain diode: Images, MAMs and MSMs
Modulated: drain ($V_D = \pm 1.2V$ pulsed)
Constant: -
Grounded: gate, source, well ($V_G = V_S = V_W = GND$)

Wavelength / Device	1319 nm NF-L-65	1064 nm NF-L-65	1319 nm PF-L-65	1064 nm PF-L-65
a) Image				
b) MAM				
c) MSM				

Table 9.4.: Reverse biased drain diode: VSs
Modulated: drain (V_D pulsed)
Constant: -
Grounded: gate, source, well ($V_G = V_S = V_W = GND$)

Wavelength / Device	1319 nm NF-L-65	1064 nm NF-L-65	1319 nm PF-L-65	1064 nm PF-L-65
d) VS	plot			
d) VS-level at V_{NOM}	D: 283 ppm G: -	D: 250 ppm G: -	D: 59 ppm G: -	D: 344 ppm G: -

drain diffusion, especially in the center of the drain. The MSM c) reveals the reason for this: there are two signals with opposite sign next to each other detectable in the center of the drain, which add up to 0 ppm in the center of the drain in the MAM. A possible origin of this signal behavior will be discussed in section 9.4.1.

d) VSs (signal-to-voltage correlation) The signal of the PF-L-65, measured with 1319 nm laser wavelength, is barely measurable - there is no signal up to -1.02 V. For the voltage sweeps, the laser was pointed to the center of the drain diffusions. As described above (in paragraph c)) in this area, there are two signals of opposite sign. So in the area, where the two signals add up to a zero ppm-value, the signal is not measurable with the voltage sweep. It is likely that the signal that was measured at voltages above -1.02 V was caused by the drift of the device, so that the laser was pointed to an area, where the signals were not adding up to zero ppm eventually. All the other signals show linear signal-to-voltage correlations.

b) + d) Signal levels (MAMs and VSs) Except the measurement of the PF-L-65 with 1319 nm laser wavelength, all the signals measured here are a factor 2.0 - 2.8 times higher than those measured on the devices of the 120 nm technology (compare to table 9.2). The ratios of the signal levels are similar to those in table 9.4: the signal levels of the NF-L-65 measurements with both laser wavelengths are in the same range, whereas the PF-L-65, measured with 1064 nm laser wavelength, has a higher signal level than the measurement with the 1319 nm laser wavelength. The PF-L-65, measured with 1319 nm laser wavelength, has a different signal behavior. The voltage sweep shows that there is no measurable signal up to -1.02 V (explanation see paragraph d)), then the signal is linear and reaches the same signal level (about 60 ppm) as the PF-L-120 (table 9.1 b) and 9.2 d)).

Summary Independent of the process technology, the modulation of the SCR thickness of the diffusion-to-well junction is the active source of the signals, so the signal contribution is the same as for the 120 nm process technology (compare to section 9.1.1). The signal-to-voltage correlation is linear as it was for the NF-L-120 and the PF-L-120 (section 9.1.1 d)). However, the differences in signal amplitude and sign compared to table 9.1 are caused by the various parameters that are changed by introducing another process technology and that are coming along with the according effects (interference, absorption and refraction).

9.1.3. Different device sizes

In this section, as an example, the results of the measurements of the PF-L-120 and the PF-S-120 - the device with the nominal gate dimensions - will be compared to discuss the signal behavior with the scaling. Note that in the case of the test structure used here, the devices with the nominal gate dimensions are built with diffusions that are lot larger than the gate - for details about layout and sizes see section 7.2. So in terms of scaling, the drains here are not as small as in a "real" device, but will show the similarities and differences between the two sizes.

Tables number 9.5 and 9.6 show the results for the two devices, PF-L-120 and PF-S-120, with the according drains driven as reverse biased diodes. The tables contain the images a), the MAMs performed at V_{NOM} (all maps set to the same ppm-scale - easier comparison) b), the respective MSMs c) and the results of the VSs of all these measurements including the maximum signal level measured at the nominal device voltage d). Both PFETs have been measured with 1319 nm and 1064 nm laser wavelength.

b) MAMs (shape of the signals) Apart from the size of the drain diffusions, the MAMs taken with the 1319 nm laser show similar, almost evenly distributed signals. The 1064 nm laser wavelength MAMs are distinct: the signal of the PF-L-120 is almost evenly distributed across the entire drain diffusion, whereas the signal of the device with the nominal gate dimensions has no signal in the center of the diffusion. This is not caused by the contrast of the reflected light (see c) and section 9.4.1).

c) MSMs (shape of the signals) The signs of the 1319 nm laser wavelength measurements are equal for both FETs: the sign in the center is negative and the diffusions are surrounded by a signal with a positive sign (sign flip due to materials in the background, see section 9.4.1). The signs of the bulk of the signals, measured with 1064 nm laser wavelength, are also negative, but there are areas in the diffusions, where the sign is positive (e.g. in the center of the drain diffusion of the device with the nominal gate dimensions). In the places, where the sign flip occurs and the positive signal interferes with the negative, in the MAMs the signal is close to zero (explanation see section 9.4.1).

d) VSs (signal-to-voltage correlation) The signal-to-voltage correlation is linear in all cases.

b) + d) Signal levels (MAMs and VSs) With the exception of the center of the PF-S-120 drain diffusion, the 1064 nm laser MAMs show higher signal levels than the 1319 nm laser MAMs. The signal levels of the 1319 nm laser measurements compare well. In the 1064 nm laser MAM, the signal level of the PF-S-120 is lower. For the voltage sweep, the laser was pointed to the center of the diffusion, where the signal was very low, due to the sign flip, but the voltage sweep shows a similar signal level as in the areas, where there was no sign flip. This might be caused by the drift of the laser to those areas during the measurement.

Summary Apart from the area, where the sign flip occurs (PF-S-120, 1064 nm laser wavelength), the signal distributions of both sizes are similar, the sign of the signals is the same and the signal levels are comparable.

Table 9.5.: Reverse biased drain diode: Images, MAMs and MSMs
Modulated: drain ($V_D = \pm 1.2V$ pulsed)
Constant: -
Grounded: gate, source, well ($V_G = V_S = V_W = GND$)

Wavelength / Device	1319 nm PF-L-120	1064 nm PF-L-120	1319 nm PF-S-120	1064 nm PF-S-120
a) Image				
b) MAM				
c) MSM				

Table 9.6.: Reverse biased drain diode: VSs
　　　　　Modulated: drain (V_D pulsed)
　　　　　Constant: -
　　　　　Grounded: gate, source, well ($V_G = V_S = V_W = GND$)

Wavelength / Device	1319 nm PF-L-120	1064 nm PF-L-120	1319 nm PF-S-120	1064 nm PF-S-120
d) VS	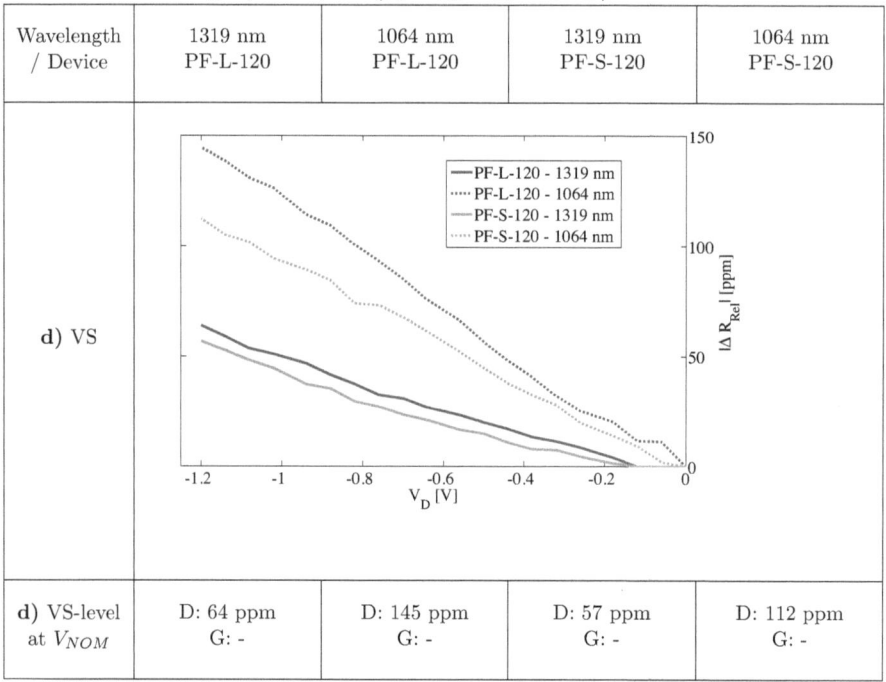			
d) VS-level at V_{NOM}	D: 64 ppm G: -	D: 145 ppm G: -	D: 57 ppm G: -	D: 112 ppm G: -

9.2. Varactor in inversion ($V_D = V_S = V_W = GND$; V_G pulsed)

For the measurements in this paragraph, only the gates of the FETs were pulsed. Drain, source and well were grounded. The PFET gates have been driven with a negative voltage, whereas the voltage for the NFET gates was positive. So the material system gate / oxide / semiconductor in the gate areas of the FETs can be understood as **MOS**-varactors in inversion (compare to section 7.4.2). For the MAMs and the MSMs, the amplitude of the voltage pulse has been set to $V_{NOM} = \pm 1.2V$. For the VSs, the laser has been pointed to the center of the gate areas and the voltage was swept linearly from 0 V up to V_{NOM}.

9.2.1. 120 nm process technology

The tables 9.7 and 9.8 show the results of the measurements for NF-L-120 and PF-L-120 with the according gates driven as varactors in inversion. Both NFET and PFET have been measured with 1319 nm and 1064 nm laser wavelength.

The tables show the images a), the MAMs performed at V_{NOM} (all maps set to the same ppm-scale - easier comparison) b), the respective MSMs c) and the results of the VSs of all these measurements including the maximum signal level measured at the nominal device voltage d).

b) MAMs (shape of the signals) The MAMs of the gate signals are similar: the complete gate areas are contributing to the signal, because - while the gates are driven with a pulse - the width of the SCR and the charge carrier density in the inversion channel underneath the gate (see description in chapter 7.4.2) are modulated. The modulation of the SCR influences the signal as described above - section 9.1.1. The inversion channel provides another layer, which contributes to the signal, because the modulation of the gate does not only vary the thickness of the SCR, but it also modulates the charge carrier density in the inversion channel, such that the refraction and absorption of the entire structure changes accordingly. So there are two possible effects - SCR and channel - that can contribute to the signal that is detected in the gate area. Here, the signals are almost evenly distributed across the gate areas, because the SCR and the inversion channel are evenly distributed along the gate area, when the drain voltage V_D is held at ground. Higher signal levels at the edges of the gates (e.g. at the source and drain sides of the NF-L-120 and PF-L-120, measured with 1064 nm laser wavelength) and in the center of the gates (see pattern in the gate areas of the 1064 nm laser wavelength measurements) are caused by the contrast of the reflected light (higher signal levels, where the image is darker).

c) MSMs (shape of the signals) The sign of the NF-L-120 signals is changing with the wavelength used (negative for 1319 nm and positive for 1064 nm), whereas the sign is the same in both cased for the PF-L-120 (negative). In opposition to the measurements of the drains driven as reverse biased diodes (see section 9.1), there is no sign flip in any

Table 9.7.: Varactor in inversion: Images, MAMs and MSMs
Modulated: gate ($V_G = \pm 1.2V$ pulsed)
Constant: -
Grounded: drain, source, well ($V_D = V_S = V_W = GND$)

Wavelength / Device	1319 nm NF-L-120	1064 nm NF-L-120	1319 nm PF-L-120	1064 nm PF-L-120
a) Image				
b) MAM				
c) MSM				

Table 9.8.: Varactor in inversion: VSs
Modulated: gate (V_G pulsed)
Constant: -
Grounded: drain, source, well ($V_D = V_S = V_W = GND$)

Wavelength / Device	1319 nm NF-L-120	1064 nm NF-L-120	1319 nm PF-L-120	1064 nm PF-L-120
d) VS	\multicolumn{4}{c}{plot: $|\Delta R_{Rel}|$ [ppm] vs V_G [V], curves: NF-L-120 - 1319 nm, PF-L-120 - 1319 nm, NF-L-120 - 1064 nm, PF-L-120 - 1064 nm}			
d) VS-level at V_{NOM}	D: - G: 107 ppm	D: - G: 167 ppm	D: - G: 129 ppm	D: - G: 122 ppm

place (the sign flipped outside the drain diffusions). Very faint signals were detected below the gate (at the bottom of the images, even to weak to be detected in the MAMs), in the area of the poly-Si line (compare to the layouts in table 7.2). This indicates that those weak signals were generated without the presence of a channel, because the only signal source in those areas can be a modulation of a SCR caused by the pulses applied to the poly-Si line (i.e. the gate).

d) VSs (signal-to-voltage correlation) The VSs show a main difference between PF-L-120 and NF-L-120. The PFET, measured with 1319 nm laser wavelength, has a linear signal-to-voltage correlation for voltages higher than -0.06 V. For the NFET, in opposition, a linear signal-to-voltage correlation has been detected in the measurements with the 1064 nm laser wavelength (at voltages higher than 0.26 V). Linear signal-to-voltage correlations were also found in the 1319 nm measurement of the NF-L-120 and the 1064 nm measurement of the PF-L-120 - at voltages higher than ± 0.6 V. Below this voltage, a sign flip occurred, which caused that the voltage sweep curve seems to have a "hump" between 0 V and ± 0.6 V (for a detailed discussion of signal contribution see section 9.4.2).

b) + d) Signal levels (MAMs and VSs) At the nominal device voltage, with a signal level of about 170 ppm in the center of the gate, the NF-L-120, measured with 1064 nm laser wavelength, has a higher level than found in the other measurements - they show comparable signal levels: NF-L-120 (1319 nm) - 107 ppm and PF-L-120 in both cases - 120-130 ppm.

Summary There are only two active signal sources of the FETs while the gates are driven as varactors in inversion: the SCR underneath the gate and the inversion channel, which are even for the case that the drain is grounded. This is the reason for the similarities in the MAMs: the signals were almost evenly distributed across the gate area. Higher signal levels at the edges of the gate or in the center of the gate were caused by the contrast of the reflected light. The signal-to-voltage correlations were partly linear in all cases. However, there were differences in the sign of the signals measured at the nominal device voltage, which were caused by interference effects due to the wavelength and the materials (discussion, see in chapter 11) and there were also sign flips in the voltage sweeps, which were caused by the electrical activity of the devices. Very faint signals were detected in the areas of the poly-Si line, at the bottom of the gates, which indicates - again - that it is possible to generate signals without the inversion channel (compare to section 9.1, SCR related signals).

9.2.2. 65 nm process technology

As in section 9.1.2, the results of the NF-L-120 and PF-L-120 will be compared with the NF-L-65 and PF-L-65. Tables number 9.9 and 9.10 show the results of the measurements for NF-L-65 and PF-L-65 with the according gates driven as varactors in inversion. Both,

NFET and PFET, have been measured with 1319 nm and 1064 nm laser wavelength.

The tables contain the images a), the MAMs performed at V_{NOM} (all maps set to the same ppm-scale - easier comparison) b), the respective MSMs c) and the results of the VSs of all these measurements including the maximum signal level measured at the nominal device voltage d).

b) MAMs (shape of the signals) The results differ a lot from those of the 120 nm process technology. The signal origin should be the same: the entire gate area contributes to the signal. But in this case, there are sign flips at the edges of the gates (towards drain and source diffusions) for all devices except the NF-L-65 measured with 1064 nm laser wavelength. So in this case, weaker or stronger signals in the gate area are caused by the interference of signals with opposite sign and not by the contrast of the reflected light - as it was the case for the 120 nm technology (see table 9.7). The edges of the gates (1064 nm laser wavelength measurements) show signal levels six times stronger than the signals in the center of the gates. The signals in the drain and source diffusions of the FETs measured with 1319 nm laser wavelength - on the other hand - are caused by the contrast of the reflected light (relatively strong signal levels in the source and drain areas, which were grounded, due to low signal levels in the static part of the reflected light, the image).

c) MSMs (shape of the signals) Especially the signals in the MSMs of the NF-L-65, measured with 1319 nm, and the PF-L-65, measured with 1064 nm, are very weak, but the signs in the center of the gates seem to be positive, except the signal of the PF-L-65, measured with 1319 nm, which is negative.

d) VSs (signal-to-voltage correlation) The signal-to-voltage correlations of the PF-L-120 and the PF-L-65 are quite similar: at 1319 nm it is linear for voltages higher than -0.32 V, whereas at 1064 nm laser wavelength a "hump" can be found as detected in the measurements in table 9.8 (120 nm technology). The signal-to-voltage correlations of the NF-L-120 and the NF-L-65 differ. Here, the correlation, measured with 1064 nm laser wavelength, is linear for voltages higher than 0.64 V, whereas the signal of the 1319 nm measurement starts linear and than saturates at voltages higher than 0.6 V. The relatively high voltage levels, from which the signals started to increase for the PF-L-65 (1319 nm) and the NF-L-65 (1064 nm), are most likely caused by the two active signal sources: the SCR underneath the gate and the inversion channel. Both signal sources either produce very low signal levels up to those voltages (PFET -0.32 V and NFET 0.64 V) or interfere and hence the signal is zero.

b) + d) Signal levels (MAMs and VSs) All the signals of the 65 nm technology are weaker than those recorded in the measurements of the 120 nm technology. Here, the NF-L-65 measurements show comparable signal levels for both wavelengths (around 70-80 ppm); the PF-L-65, measured with 1064 nm laser wavelength, has a lower signal level

Table 9.9.: Varactor in inversion: Images, MAMs and MSMs
  Modulated: gate ($V_G = \pm 1.2V$ pulsed)
  Constant: -
  Grounded: drain, source, well ($V_D = V_S = V_W = GND$)

Wavelength / Device	1319 nm NF-L-65	1064 nm NF-L-65	1319 nm PF-L-65	1064 nm PF-L-65
a) Image				
b) MAM				
c) MSM				

Table 9.10.: Varactor in inversion: VSs
Modulated: gate (V_G pulsed)
Constant: -
Grounded: drain, source, well ($V_D = V_S = V_W = GND$)

Wavelength / Device	1319 nm NF-L-65	1064 nm NF-L-65	1319 nm PF-L-65	1064 nm PF-L-65
d) VS	\multicolumn{4}{c}{plot of $|\Delta R_{Rel}|$ [ppm] vs V_G [V]}			
d) VS-level at V_{NOM}	D: - G: 79 ppm	D: - G: 72 ppm	D: - G: 104 ppm	D: - G: 60 ppm

than measured with 1319 nm - 60 ppm compared to 104 ppm.

Summary The gate signals recorded here differ a lot from those of the 120 nm technology. The MAMs detected signals at the edges of the gates with increased or decreased signal levels, which are not caused by the contrast of the reflected light, but by a sign flip across the gate area (sign flip from the inner area of the gate to the edges - compare to the MSMs c)). The signal-to-voltage correlations of the PF-L-65 are comparable to those of the PF-L-120; the correlations of both NFETs are distinct. All signal levels detected here are lower than those of the 120 nm technology.

9.2.3. Different device sizes

As in section 9.1.3, the results of the measurements of the PF-L-120 and the PF-S-120 - the device with the nominal gate dimensions - will be compared to discuss the signal behavior with the scaling. In opposition to the diffusion sizes, the gate dimensions are indeed of nominal size - see section 7.2.

Tables number 9.11 and 9.12 show the results for the two devices PF-L-120 and PF-S-120 with the according gates driven as varactors in inversion. The table contains the images a), the MAMs performed at V_{NOM} (all maps set to the same ppm-scale - easier comparison) b), the respective MSMs c) and the results of the VSs of all these measurements including the maximum signal level measured at the nominal device voltage d). Both PFETs have been measured with 1319 nm and 1064 nm laser wavelength.

b) MAMs (shape of the signals) In the MAMs of the PF-S-120 no signals are detectable. The results of the PF-L-120 were discussed in section 9.2.1.

c) MSMs (shape of the signals) The signals in the MSMs of the PF-S-120 are very faint along the entire poly-Si line - not only in the gate area, see section 7.2 for the layout. The measurements with the 1319 nm laser indicate negative signs for both PFETs, whereas the sign of the 1064 nm laser measurements differ from each other: the sign measured at the PF-L-120 was negative; the PF-S-120 shows a very faint positive signal.

d) VSs (signal-to-voltage correlation) and b) + d) signal levels (MAMs and VSs) As in the MAMs, performed for the PF-S-120, the VSs show no signals - only noise detectable. The maximum noise level measured with 1064 nm laser wavelength was fairly high - up to 23 ppm - but a correlation between signal and voltage could not be extracted. For the discussion of the results of the PF-L-120 see section 9.2.1.

Summary When only the gates of the FETs are driven as varactors in inversion, the size of the gates is vital for the strength of the signal that is been detected. The PF-S-120 shows no signals in the MAMs or in the VSs. Very faint signals were only detected in the

Table 9.11.: Varactor in inversion: Images, MAMs and MSMs
Modulated: gate ($V_G = \pm 1.2V$ pulsed)
Constant: -
Grounded: drain, source, well ($V_G = V_S = V_W = GND$)

Wavelength / Device	1319 nm PF-L-120	1064 nm PF-L-120	1319 nm PF-S-120	1064 nm PF-S-120
a) Image				
b) MAM				
c) MSM				

Table 9.12.: Varactor in inversion: VSs
  Modulated: gate (V_G pulsed)
  Constant: -
  Grounded: drain, source, well ($V_D = V_S = V_W = GND$)

Wavelength / Device	1319 nm PF-L-120	1064 nm PF-L-120	1319 nm PF-S-120	1064 nm PF-S-120
d) VS				
d) VS-level at V_{NOM}	D: - G: 129 ppm	D: - G: 122 ppm	D: - G: only noise	D: - G: only noise

MSMs; not only in the gate areas, but along the entire poly-Si line, which indicates that the signal is generated by the SCR underneath the poly-Si (compare to section 9.2.1).

9.3. FET, gate and drain pulsed ($V_S = V_W = GND$; $V_G = V_D$ pulsed simultaneously)

For the measurements in this paragraph, both - gates and drains - of the FETs were pulsed simultaneously. Source and well were grounded. The PFETs have been driven with a negative voltage, whereas the voltage for the NFETs was positive. In this case, the FET is switching between off-state and on-state, in which the current is flowing from source to drain (compare to section 7.4.3). For the MAMs and the MSMs, the amplitude of the voltage pulse has been set to the nominal device voltage ($V_{NOM} = \pm 1.2V$). For the VSs the laser has been pointed to the center of the gate and drain diffusions and the voltage was swept linearly from 0 up to the nominal device voltage.

9.3.1. 120 nm process technology

Tables number 9.13, 9.14 and 9.15 show the results of the measurements for NF-L-120 and PF-L-120 with the according drains and gates driven simultaneously. Both, NFET and PFET, have been measured with 1319 nm and 1064 nm laser wavelength.

The tables contain the images a), the MAMs performed at V_{NOM} (all maps set to the same ppm-scale - easier comparison) b), the respective MSMs c) and the results of the VSs of all these measurements including the maximum signal level measured at the nominal device voltage d). The VSs were performed twice for each device (one time, with the laser pointed to the center of the drain (compare to section 9.1.1) and the second time with the laser pointed to the center of the gate (compare to section 9.2.1); the results are shown separately in the table.

b) MAMs (shape of the signals) **Drain signals:** The MAMs of the drain areas show a very similar signal distribution as detected in the measurements of the reverse biased diodes (compare to table 9.1). **Gate signals:** The signals in the gate areas are no longer evenly distributed: in all cases, the signal decreases along the channel towards the drain. While the FETs are driven with the nominal device voltage, as it was the case for the MAMs, the thickness of the SCR underneath the gate is increasing towards the drain and the thickness of the inversion channel is decreasing towards the drain (both were evenly distributed for the varactors in inversion). These two layers are influencing the reflection of the structure due to their optical properties (absorption, refraction, thickness of the layer), so there are two possible signal sources that are contributing to the signal and that explain the differences in amplitude, signal shape and sign of the signals along the channel - see section c)). A discussion of the signal contribution can be found in section 9.4.

Table 9.13.: FET, gate and drain pulsed simultaneously: Images, MAMs and MSMs
Modulated: gate and drain ($V_G = V_D = \pm 1.2V$ pulsed)
Constant: -
Grounded: source, well ($V_S = V_W = GND$)

Wavelength / Device	1319 nm NF-L-120	1064 nm NF-L-120	1319 nm PF-L-120	1064 nm PF-L-120
a) Image				
b) MAM				
c) MSM				

Table 9.14.: NF-L-120, gate and drain pulsed simultaneously: VSs
Modulated: gate and drain ($V_G = V_D$ pulsed)
Constant: -
Grounded: source, well ($V_S = V_W = GND$)

Wavelength / Device	1319 nm NF-L-120	1064 nm NF-L-120
d) VS in the center of the drain	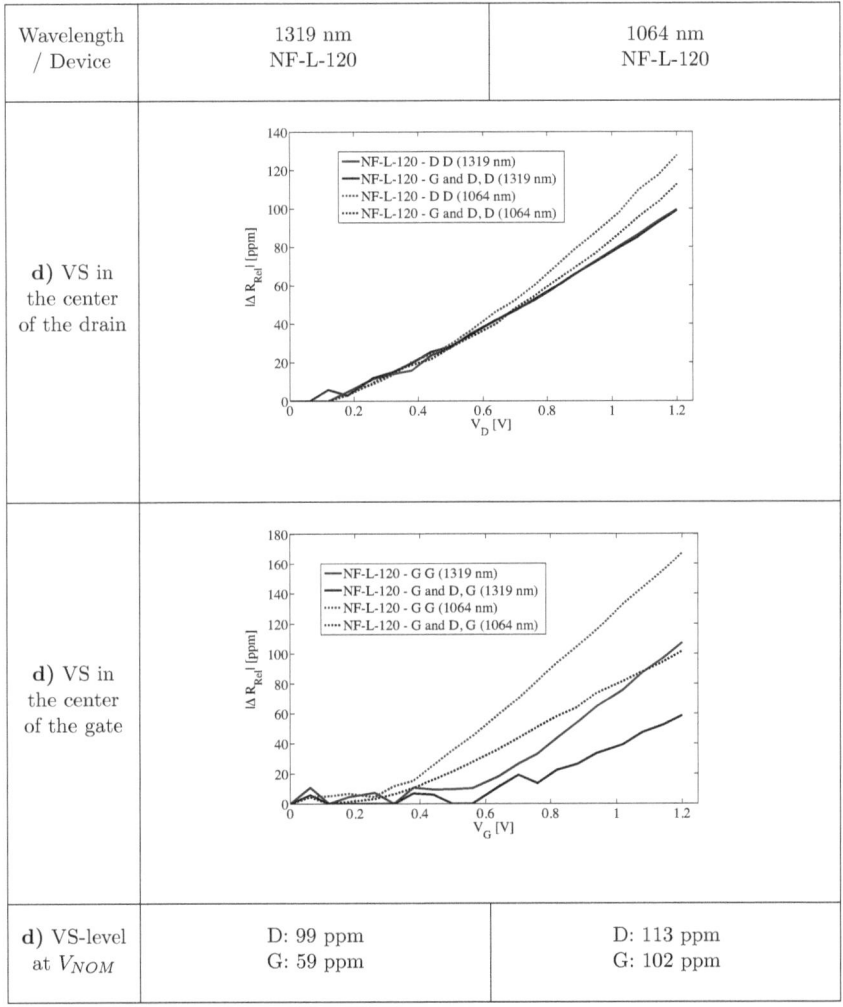	
d) VS in the center of the gate		
d) VS-level at V_{NOM}	D: 99 ppm G: 59 ppm	D: 113 ppm G: 102 ppm

Table 9.15.: PF-L-120, gate and drain pulsed simultaneously: VSs
Modulated: gate and drain ($V_G = V_D$ pulsed)
Constant: -
Grounded: source, well ($V_S = V_W = GND$)

Wavelength / Device	1319 nm PF-L-120	1064 nm PF-L-120
d) VS in the center of the drain	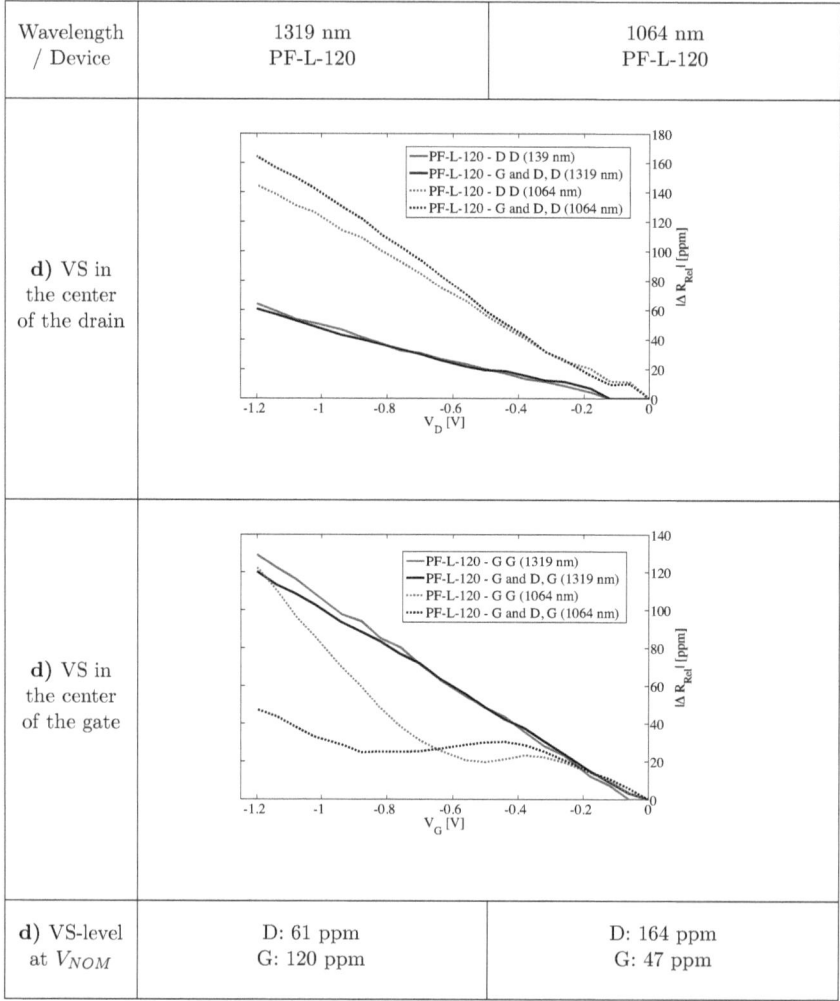	
d) VS in the center of the gate		
d) VS-level at V_{NOM}	D: 61 ppm G: 120 ppm	D: 164 ppm G: 47 ppm

c) MSMs (shape of the signals) **Drain signals:** The MSMs show the same signs in the drain areas as detected in the measurements of the diodes in reverse bias (compare to table 9.1). **Gate signals:** The sign of the gate signals close to the source is the same as measured in section 9.2.1. The NF-L-120, measured with 1319 nm, and the PF-L-120, measured with 1064 nm laser wavelength, show signals of opposite sign in the gate area close to the drain in comparison to the source side. This opponent signal is causing the decrease of the signals in the MAMs in the gate area. For the other two cases (NF-L-120 (1064 nm) and PF-L-120 (1319 nm)), the decrease of the signal along the channel should have the same reason, but the signal, which is decreasing the signal along the channel, is too weak to be detected.

d) VSs (signal-to-voltage correlation) **Drain signals:** The signal-to-voltage correlations of the drain signals are also quite similar to those measured in section 9.1.1. The signals in the VSs here show marginal lower or higher signal levels, but this is caused by the slight variations of the probe placement. **Gate signals:** The signal-to-voltage correlation of the gate signals is also similar to the results shown in table 9.8. Lower signal levels were detected with the voltage sweep, due to the probe placement: the laser has been pointed to the center of the gates and in that area the signal is lower than in the according cases of the varactor in inversion (compare to table 9.7) due to the signal decrease along the channel. The PF-L-120, measured with 1319 nm, has not such a strong signal decrease along the gate as all the other FETs. The maximums of the "humps" that were measured with the 1064 nm laser in section 9.2.1 are shifted to different voltage levels - from 0.18 V to 0.06 V for the NF-L-120 (to lower levels) and from -0.38 V to -0.44 V for the PF-L-120 (to higher levels).

b) + d) Signal levels (MAMs and VSs) **Drain signals:** The signal levels of the drain areas are comparable to those detected in the measurements of the diodes in reverse bias - compare to section 9.1 and 9.2. **Gate signals:** The signal levels along the channel decrease towards the drain due to the two signals generated by the gate: space charge region and inversion channel.

Summary The signs, amplitudes, signal-to-voltage correlations and distributions of the drain signals are very similar to those of the reverse biased diode measurements - see section 9.1.1. The gate signals close to the source show the same sign and amplitude as the varactors in inversion, but the signal decreases towards the drain due to a sign flip that has been detected in two MSMs. This effect also influences the signal-to-voltage correlations.

9.3.2. 65 nm process technology

Tables 9.16, 9.17 and 9.18 show the results of the measurements for NF-L-65 and PF-L-65 with the according drains and gates pulsed simultaneously. Both, NFET and PFET, have been measured with 1319 nm and 1064 nm laser wavelength.

The tables contain the images a), the MAMs performed at V_{NOM} (all maps set to the same ppm-scale - easier comparison) b), the respective MSMs c) and the results of the VSs of all these measurements including the maximum signal level measured at the nominal device voltage d). As in section 9.3.1, the VSs were performed twice for each device - in the center of gate and drain - and the results are shown separately in the table.

b) MAMs (shape of the signals) **Drain signals:** The drain signal distributions and amplitudes are very similar to those detected in the measurements of the reverse biased diodes in section 9.1.2. **Gate signals:** The signals close to the source are comparable to the measurements of the varactors in inversion in section 9.2.2. In comparison to those levels, the gate signals close to the drain are increased in the 1319 nm laser measurements and decreased in the measurements performed with the 1064 nm laser.

c) MSMs (shape of the signals) **Drain signals:** The signs of the drain signals do not differ from those detected in the measurements of the diodes in reverse bias in section 9.1.2. **Gate signals:** At the source side of the gate, the sign of the signals is the same as in the measurements of the varactors in inversion (compare to table 9.9). The signs of the signals at the drain side are the opposite of those detected in the varactor measurements (table 9.9). This explains the signal decrease along the channel (see also section 9.2.2).

d) VSs (signal-to-voltage correlation) **Drain signals:** The signal-to-voltage correlations of the drain signals in section 9.1.2 and here are much alike. The only exception is the PF-L-65, measured with 1319 nm laser wavelength: the signal shown in section 9.1.2 was very noisy and weak, but eventually the VS measured a linear signal-to-voltage correlation. Here, the signal is similar - partly linear (for voltages higher than -0.56 V) -, but has a higher amplitude. Most likely, the difference between these two signals is caused by the probe placement. **Gate signals:** The signal-to-voltage correlations of the gate signals seem to be similar; the signal levels are different (explanation see c)).

b) + d) Signal levels (MAMs and VSs) **Drain signals:** The drain signal levels are similar to those in the measurements of the diodes in reverse bias (see section 9.1.2). The slight differences in the signal levels are most likely caused by the probe placement. **Gate signals:** Except the gate signal of the NF-L-65, measured with 1064 nm laser wavelength, the signals here are higher than detected in the measurements of the varactors in inversion (in section 9.2.2).

Summary The signs, amplitudes, signal-to-voltage correlations and distributions of the drain signals are very similar to the measurements of the reverse biased diodes (section 9.1.2). The gate signals close to the source have the same sign and amplitude as the measurements of the varactors in inversion (see section 9.2.2), but the signal decreases towards the drain due to a sign flip that has been detected in the MSMs. This effect also influences the signal-to-voltage correlations.

Table 9.16.: FET, gate and drain pulsed simultaneously: Images, MAMs and MSMs
Modulated: gate and drain ($V_G = V_D = \pm 1.2V$ pulsed)
Constant: -
Grounded: source, well ($V_S = V_W = GND$)

Wavelength / Device	1319 nm NF-L-65	1064 nm NF-L-65	1319 nm PF-L-65	1064 nm PF-L-65
a) Image				
b) MAM				
c) MSM				

Table 9.17.: NF-L-65, gate and drain pulsed simultaneously: VSs
Modulated: gate and drain ($V_G = V_D$ pulsed)
Constant: -
Grounded: source, well ($V_S = V_W = GND$)

Wavelength / Device	1319 nm NF-L-65	1064 nm NF-L-65
d) VS in the center of the drain	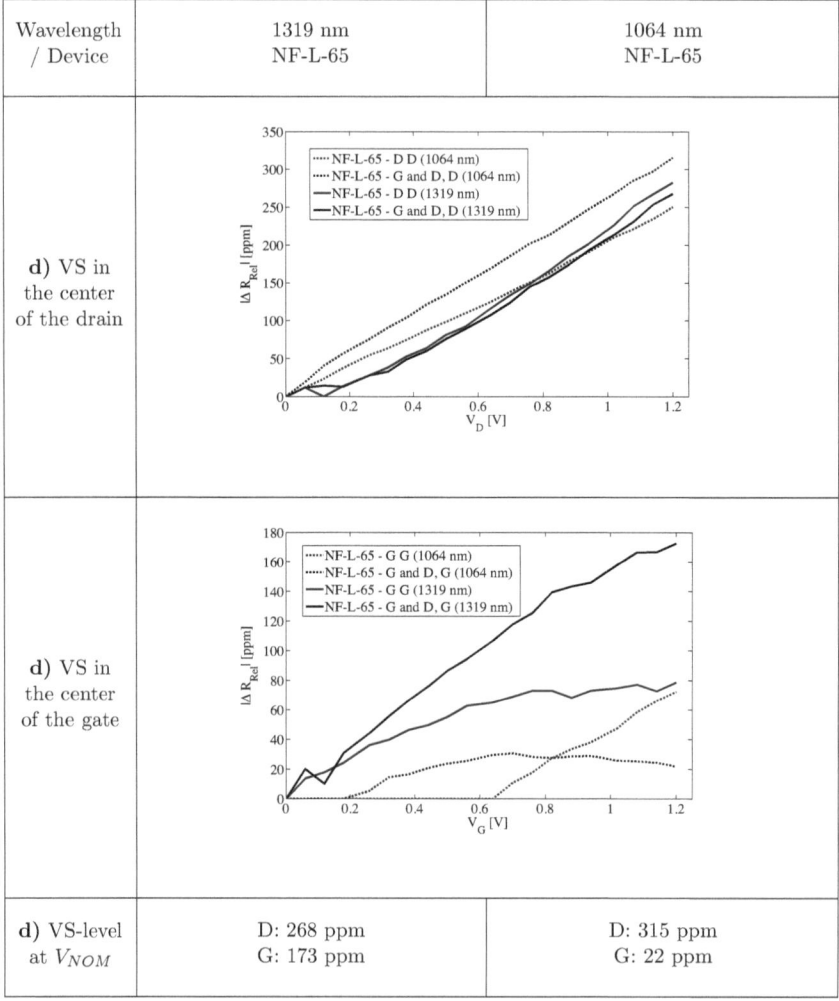	
d) VS in the center of the gate		
d) VS-level at V_{NOM}	D: 268 ppm G: 173 ppm	D: 315 ppm G: 22 ppm

Table 9.18.: PF-L-65, gate and drain pulsed simultaneously: VSs
Modulated: gate and drain ($V_G = V_D$ pulsed)
Constant: -
Grounded: source, well ($V_S = V_W = GND$)

Wavelength / Device	1319 nm PF-L-65	1064 nm PF-L-65
d) VS in the center of the drain	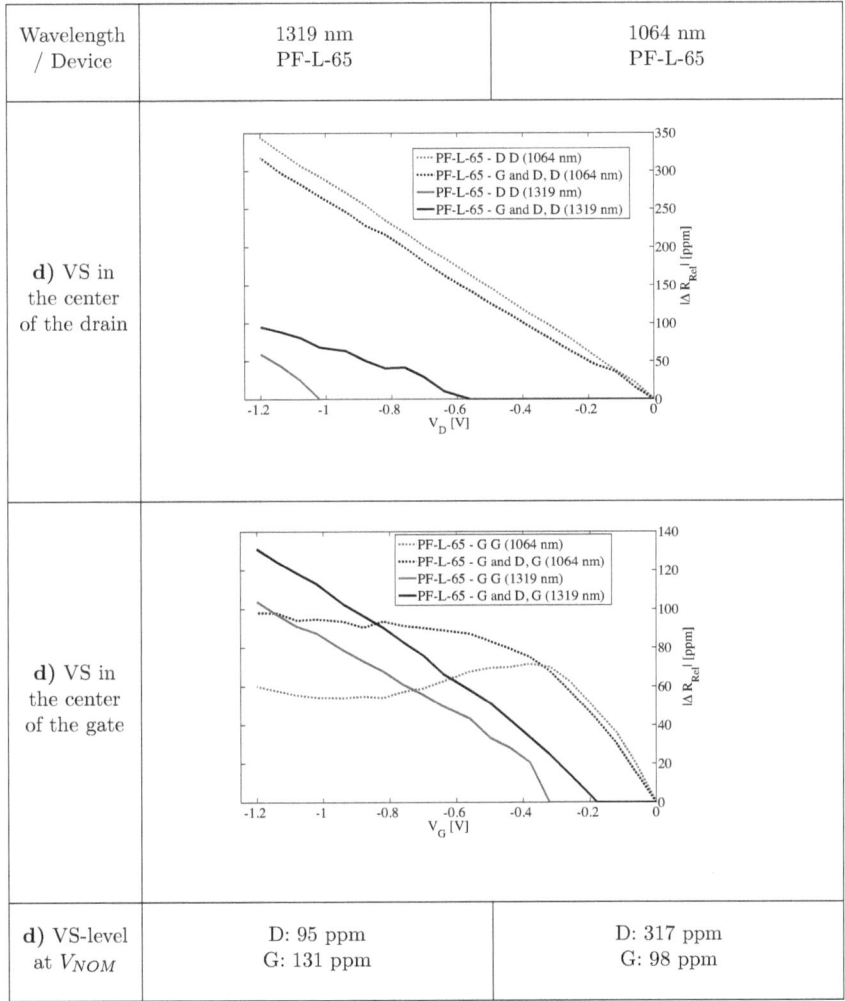	
d) VS in the center of the gate		
d) VS-level at V_{NOM}	D: 95 ppm G: 131 ppm	D: 317 ppm G: 98 ppm

9.3.3. Different device sizes

In this section, as an example, the results of the measurements of the PF-L-120 and the PF-S-120 - the device with the nominal gate dimensions - will be compared to discuss the signal behavior with the scaling.

Tables number 9.19 and 9.20 show the results for the two devices PF-L-120 and PF-S-120 with the according drains and gates driven simultaneously. The tables contain the images a), the MAMs performed at V_{NOM} (all maps set to the same ppm-scale - easier comparison) b), the respective MSMs c) and the results of the VSs of all these measurements including the maximum signal level measured at the nominal device voltage d). Both PFETs have been measured with 1319 nm and 1064 nm laser wavelength.

b) MAMs (shape of the signals) **Drain signals:** The signal levels and distributions of the drain signals here and those of the reverse biased diodes (tables 9.5 and 9.6) are much alike. **Gate signals:** The gate signals of the 10 µm structures were discussed above. In opposition to the varactor in inversion, the gate of the PF-S-120 here shows a strong signal (compare to table 9.11, where the entire poly-Si line produced a weaker signal).

c) MSMs (shape of the signals) **Drain signals:** As the MAMs, the MSMs show no differences compared to the drains driven as reverse biased diodes (see section 9.1.3). **Gate signals:** The sign of the PF-S-120, measured with 1064 nm laser wavelength, is negative. Assuming that the stronger signal in the gate area dominates the overall signal in the gate area, this signal sign can be derived from the downscaling of the 10 µm device: the sign of the stronger signal in the gate of the PF-L-120 is negative, too. The signal of the PF-S-120, measured with 1319 nm laser wavelength, is so weak that, in the MSM, it can hardly be distinguished from the drain signal. However, the overall sign needs to be positive in this case, because a negative sign would have decreased the amplitude of the ring shaped signal surrounding the drain in the MAM. The stronger signal in the gate area of the PF-L-120 had a negative sign, which means that the stronger signal does not dominate the overall sign in this case, when the device is scaled down. The signal with positive sign was not detected with the MSM tool, but was derived from the decrease of the signal along the channel.

d) VSs (signal-to-voltage correlation) **Drain signals:** Especially for the PF-S-120, measured with 1064 nm laser wavelength, the signal level is strongly depending on the probe placement (see discussion of the drain diodes in reverse bias, tables 9.5 and 9.6), but basically the drain signals show the same signal-to-voltage correlations as the drains driven as diodes in reverse bias. **Gate signals:** The signal amplitude and signal-to-voltage correlation of the PF-L-120 depends on the probe placement. For the PF-S-120, the probe placement is even more important, because the signal is not measurable, if the laser is pointed to the wrong position. The VS shown in table 9.20 has a parabolic signal-to-voltage correlation.

Table 9.19.: FET, gate and drain pulsed simultaneously: Images, MAMs and MSMs
Modulated: gate and drain ($V_G = V_D = \pm 1.2V$ pulsed)
Constant: -
Grounded: source, well ($V_S = V_W = GND$)

Wavelength / Device	1319 nm PF-L-120	1064 nm PF-L-120	1319 nm PF-S-120	1064 nm PF-S-120
a) Image				
b) MAM				
c) MSM				

Table 9.20.: PF-S-120, gate and drain pulsed simultaneously: VSs
Modulated: gate and drain ($V_G = V_D$ pulsed)
Constant: -
Grounded: source, well ($V_S = V_W = GND$)

Wavelength / Device	1319 nm PF-S-120	1064 nm PF-S-120
d) VS in the center of the drain	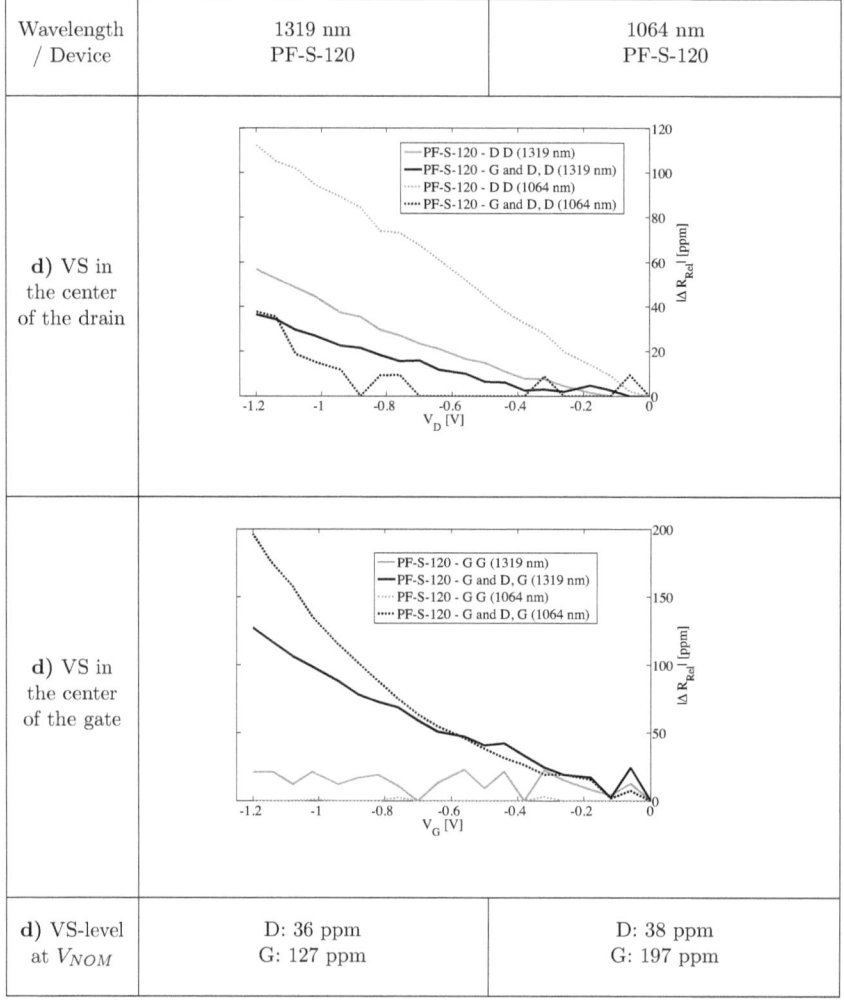	
d) VS in the center of the gate		
d) VS-level at V_{NOM}	D: 36 ppm G: 127 ppm	D: 38 ppm G: 197 ppm

b) + d) Signal levels (MAMs and VSs) **Drain signals:** As discussed in paragraph c) and d) above, the signal levels are comparable to the measurements of the reverse biased diodes (PF-S-120, measured with 1064 nm: signal level strongly depending on the probe placement). **Gate signals:** For both FETs, the signal levels are strongly depending on the probe placement: due to the signal decrease along the channel for the PF-L-120 and due to the small-sized signal area for the PF-S-120. The VS for the PF-S-120 has been performed several times, to verify that the gate signal is measurable. Eventually it was possible to place the probe to a position with similar signal level as measured in the MAM: the according VS is shown in table 9.20. The signal level at V_{NOM}, measured with the VS, reached 197 ppm.

Summary Basically, the signs, amplitudes and signal-to-voltage correlations of the drain signals here are very similar to those of the diodes in reverse bias (compare to section 9.1.3). The slight differences in the signal amplitude are caused by the probe placement. The gate signal of the PF-L-120 decreased along the channel, when both pins are pulsed. The gate signals of the PF-S-120 are only detectable, when both pins are pulsed; the gate signals then become much stronger than the according drain signals (see table 9.19). Signal-to-voltage correlations and signal amplitudes of the device with the nominal gate dimensions indicate that interference of gate and drain signals take place. The signs of the PFET gate signals, measured with 1064 nm laser wavelength, seem to be related: the sign of the stronger signal in the gate of the PF-L-120 dominates, when the device is scaled down in size, so the sign of the small structure is negative, too. This is not the case for the 1319 nm laser wavelength measurement.

9.4. Signal contribution

In the previous sections, several differences between the signals of the devices were recorded: various signal-to-voltage correlations, signal levels, shapes of the signals in the MAMs and signs were detected. This section will point out the signal contribution of areas, where the sign flips, it will explain gate and drain signals and discuss the reason for the broad spectrum of results for the devices used.

9.4.1. Signal contribution - sign flips (MSMs) and signal level variations (MAMs) within the same area

This paragraph will discuss the origin of the sign flip and the according signal levels in the MAMs that were detected in several measurements. As examples, NF-S-120 and PF-S-120 will be discussed briefly.

Figure number 9.1 (b) and c)) shows the MAMs of these FETs, the according drains driven as reverse biased diodes, measured with 1064 nm laser wavelength. The layout and an overlay of layout and image are shown, too (a)).

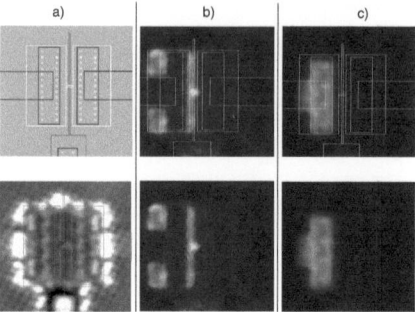

Figure 9.1.: a) Layout and overlay of layout and image for P- and NF-S-120 b) MAM of PF-S-120 and c) MAM of NF-S-120, drains driven as reverse biased diodes, measured with 1064 nm laser wavelength.

The signals in the MAMs are the complement of each other. The shape of the signal is correlated to the shape of metal 1. In the MAMs, the signal is close to zero within this area (PF-S-120, figure 9.1 b)) or outside of this area (NF-S-120, figure 9.1 c)). In the MSMs (not shown in the figure), the signal is spread across the entire diffusion areas, but the sign of the signal in the area with the metalization in the back (PF-S-120: positive; NF-S-120: negative) is the opposite of the signal outside of this area (PF-S-120: negative; NF-S-120: positive).

This sign flip occurs, because the metal is highly reflective and changes the phase of the signal by 180°. The signal in the outer area, where there is no metal in the background, is most likely the signal that is caused by the SCR modulation. The signal development for the area with the metal in the background can only be explained in a very simple way by introducing an amplification factor for the reflection. Assuming that the modulation of the SCR causes a signal of e.g. 150 ppm in both cases (positive sign for the NFET and negative sign for the PFET) the calculations below only agree with the recorded signals, if the amplification factors are chosen to be 1.333 for the PFET and 3 for the NFET. The differences in the amplification factors could be explained with interference effects caused by different thicknesses of the according layers (diffusion, SCR etc.) of the two FETs.

PFET:

1. Signal caused by the modulation of the SCR (PFET):
   -150 ppm

2. Changed signal due to reflection off interconnect: 180° phase shift and factor 1.333 amplification:
   +200 ppm

3. Resulting signal (interference of 1. and 2.; simple addition):
   +50 ppm

NFET:

1. Signal caused by the modulation of the SCR (NFET):
   +150 ppm

2. Changed signal due to reflection off interconnect: 180° phase shift and factor 3 amplification:
   -450 ppm

3. Resulting signal (interference of 1. and 2.; simple addition):
   -300 ppm

At this point it becomes clear that - especially for smaller structures - it is very important, where the laser is pointed during measurements. To detect a signal, in this case, it is vital to use the MAM tool, because the VS and the time-domain measurement might just miss the right spot. Especially the MSM is very useful, because it is possible to detect even very faint signals, caused e.g. by a sign flip, due to the higher sensitivity of the modulation sign mapping tool (see chapter 6.2.2).

Signal generation effects - as just outlined -, where the signal is increased or decreased due to the material properties in the background and the interference effects that come along, are omnipresent in the devices. All signals that were increased or decreased compared to the signals in the exact same measurement and in the same area (except the signal decrease along the channel, when both pins are pulsed) can be traced back to this effect. Examples for such signals are:

- the pattern in the gate areas, which were detected in the MAMs of the PF-L-120 and NF-L-120, measured with 1064 nm laser wavelength,

- the signal decrease in the MAMs of the inner part of the drain diffusions, measured with 1064 nm laser wavelength,

- and the signal decrease (MAMs) and the sign flip (MSMs) of all the signals surrounding the diffusions.

Note that the sign does not necessarily have to flip at one of the according interfaces to influence the signal strength.

9.4.2. Signal contribution - sign flip or signal amplitude variations in the VSs, signal level variations along the gate (gate signal)

Several VSs of gate areas revealed "hump"-shaped signal-to-voltage correlations (e.g. NF-L-120, measured with 1319 nm, PF-L-120 measured with 1064 nm). These signal-to-voltage correlations come along with a sign flip (this was proven with time-domain measurements). Since those signals were detected with the VS, it is not the overall sign of the signals at all voltages - as discussed in section 9.1 -, but a sign changing with the applied gate voltage. Since there are two active signal sources, when the gate is pulsed - SCR and inversion channel -, it is possible that the two effects are causing signals of opposite signs (not necessarily, hence not all gate signals showed "humps"). This means that there are two competing signals that are of different strengths - depending on the voltage that is applied - and the one that is stronger prevails in the measurement and even determines the sign at the nominal device voltage in the MSMs.

Since the strength of the two signal sources - SCR and inversion channel - also depends on the position in the device (compare to figure 7.1), this also explains the sign flips in the gate areas found in the MSMs and the signal decrease in the MAMs along the gate of several devices, when both, gate and drain, are pulsed. The sign of the NF-L-120 signal with both pins pulsed, measured with 1319 nm laser wavelength, see table 9.13, flipped in the gate area (MSM), the signal in the MAM decreased towards the drain and the VS in the center of the gate showed a "hump"-shaped signal-to-voltage correlation, as described in the paragraph above. Figure number 9.2 contains the VSs for three laser positions in the gate area of the NF-L-120: close to the drain, close to the source and in the center of the gate (the latter is identical to that shown in table 9.14). The signal-to-voltage correlations of the three positions are distinct, which confirms the variations of signal strength of the two signal sources along the gate. It can be assumed that the signal that was measured close to the drain represents the SCR-signal source. Close to the source the inversion channel has a strong influence on the signal and in the center of the gate, both signal sources influence the overall behavior. The signal levels at the nominal device voltage represent the levels of the MAM of the according positions (note that the signal level close to the source is comparable to that of the varactor in inversion, compare to table 9.8).

9.4.3. FET signal contribution

In several measurements, strong signals were detected, only when gate and drain are pulsed at the same time (compare the measurements of the device with the nominal gate dimensions in sections 9.2.3 and 9.3.3). It is not likely that the signal strength is correlated to the *current* density, because an LVP signal is generated by the free carriers and not necessarily by the current of a device. However, in the following, the current densities are estimated very roughly. The results can be found in table 9.21 (A is the

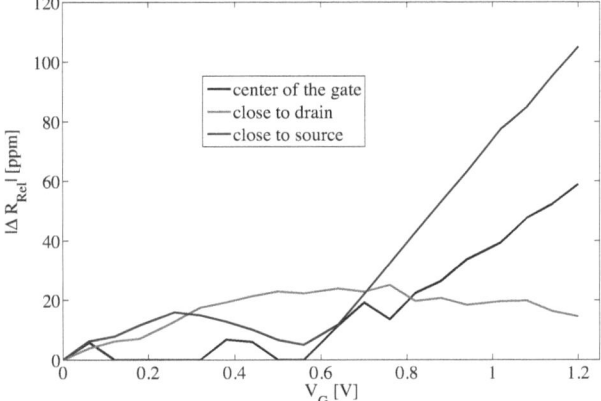

Figure 9.2.: VSs of NF-L-120, gate and drain pulsed simultaneously, three positions in the gate area: close to the drain, close to the source and in the center of the gate

channel cross-section [1], i.e. the area the current passes). As expected, the table shows that the signal strength is not directly correlated to the current density. If the signal strength would increase linearly with the current density, it meant that the PFET signal was lower than the NFET signal (comparable only among devices of the same process technology). And it meant that the device with the nominal gate dimensions had the highest signal level. But the signal levels of the PF-L-120 were higher than the levels of the NF-L-120 - see table 9.13. The highest signal levels have not been detected in the measurements of the PF-S-120, but this might be caused by the resolution or interference effects (obliteration of the overall signal due to opposite signs of gate and drain signals) coming along with downscaling of a structure to sub-micron dimensions.

9.4.4. Signal contribution summary

Table 9.22 contains the MAMs and the MSMs of the 1064 nm measurements - as an example - of the PF-L-120 and the NF-L-120. The device modi used here are as follows:

1. The varactors in inversion - device modi and signal contribution as in section 9.2.

---

[1] A is the area (cross-section), where the current flows through - for example at the source side. Rough estimation: if the diffusions, the wells, the gates and the applied voltages are all the same (without any drain voltage applied), then the channel thicknesses of the three devices should be the same in all cases, too. Here, 3 nm is a rough estimation.

Table 9.21.: Estimation of current densities: PF-L-120, NF-L-120 and PF-S-120

Device	W [μm] / L [μm]	Channel cross-section $A = W \cdot 3$nm [m^2]	On current I_{ON} [μA]	Current density $J = \frac{|I_{ON}|}{A}$ [$\frac{A}{m^2}$]
PF-L-120	10 / 10	$3 \cdot 10^{-14}$	-31	$1 \cdot 10^9$
NF-L-120	10 / 10	$3 \cdot 10^{-14}$	150	$5 \cdot 10^9$
PF-S-120	0.16 / 0.12	$4.8 \cdot 10^{-16}$	-30	$6.3 \cdot 10^{10}$

2. The drains were pulsed with the nominal device voltage and the gates were held constant, also at the nominal device voltage (high), so the FETs are in the same state as if both pins of the FETs were pulsed, only that the gate is not modulated and thus not contributing to the modulated signal itself.

3. FETs with both pins pulsed - device modi and signal contribution as in section 9.3.

Drain signals

When the drains of the FETs were pulsed - no matter, which other pins were also used - the drain diffusions always showed the same signal (compare the MAMs and MSMs of 1) and 3) in table 9.22). Not only the sign of the signals stayed the same, but also the shape of the signal in the MAMs, and the signal-to-voltage correlations (all signals were at least partly linear) were quite similar. Smaller variations in the signal levels and slightly different signal-to-voltage correlations were caused by the probe placement. The sources of these signals were the SCRs underneath the drains.

Gate signals

In the contrary, the signals that were detected in the gate areas changed with different device modi and showed a much higher complexity in the overall signal contribution. The following paragraphs will explain the signal contribution of the gate signals in general.

Table 9.22.: Signal contribution: NF-L-120 and PF-L-120, measured with 1064 nm laser wavelength
1) Varactor in inversion ($V_G = \pm 1.2V$ pulsed; $V_D = V_S = V_W = GND$)
2) FET, drain pulsed, gate high ($V_D = \pm 1.2V$ pulsed; $V_G = \pm 1.2V$ constant; $V_S = V_W = GND$)
3) FET, gate and drain pulsed simultaneously ($V_G = V_D = \pm 1.2V$ pulsed; $V_S = V_W = GND$)

Device	1) Varactor in inversion	2) FET, drain pulsed, gate high	3) FET, gate and drain pulsed
NF-L-120 MAM			
NF-L-120 MSM			
PF-L-120 MAM			
PF-L-120 MSM			

1) Varactors in inversion

As explained in section 9.2, the signal origin of the gate signals, when only the gate is pulsed, is the SCR underneath the gate and the inversion channel. Both are evenly distributed along the gate. As can be found in table 9.22 1) and section 9.2, these two signal sources can result in various signal-to-voltage correlations, signal strengths and a negative but also a positive sign - depending on the interference effects that come along with the types of FETs, the size, the process technology and the laser wavelength used for the measurement.

2) FET, drain pulsed, gate high

When the gate voltage is held constant and the drain is pulsed (see table 9.22 2)), the signal detected in the gate areas can only be caused by the modulation of the drain, because the voltage at the drain is the only voltage that is modulated (gate is static). So the signals in the gate areas shown in table 9.22 2) are caused by the influence of the drain voltage on the channel and / or the SCR underneath the gate. The effects that occur are as follows: if a pulse is applied to the drain, the thickness of the SCR underneath the gate close to the drain increases and the thickness of the inversion channel close to the drain decreases.

3) FET, gate and drain pulsed

In this case both pins, gate and drain, are pulsed with the nominal device voltage, so the signal contribution is a composition of the drain signal, caused by the SCR and the gate signal, which is caused by the inversion channel and the SCR underneath the gate plus the influence of the drain on the latter two. So the overall signal can be understood as a composition of the varactor in inversion (table 9.22 1)) and the influence on this signal caused by the drain modulation (table 9.22 2)). This produces a signal in the drain area, which is basically like the drain signals as detected for the reverse biased diodes, and a signal in the gate area that is comparable to the varactor in inversion with subtraction of the drain influence (table 9.22 3)).

Sub-micron devices

Especially the measurements that were performed on the devices with the nominal gate dimensions showed that the signal strength is the highest, when the two effects of the drain and the gate - SCR underneath the drain / gate and inversion channel - occur simultaneously. It is not clear, if this is caused by the electrical effects (e.g. SCR or inversion charge carrier density) that influence the optical properties of the structure or by the interference effects - e.g. the gate signal simply adds / enhances the drain signal, when both signals interfere, because they are closer together. The signal-to-voltage correlations, however, indicate that drain and gate signal interfere and both contribute to the signal.

It is likely that the resolution is also important for the signal strength. If the 10 μm structures were shrunk down in size, maybe the signal would increase, just because parts of the signals are superimposing (interfering). Shrinking even further would mean that the detection scheme was not longer able to catch all the reflected light (resolution), which might cause a signal that is even weaker.

9.5. Evaluation of the measurement methods

Conventionally, tools used the time-domain measurement method to extract waveforms from a device (compare to section 6.2.1). For the measurements of this work, frequency-domain measurements were used to obtain information about signs, amplitudes and signal-to-voltage correlations.

With the MAMs and the MSMs it is possible, to scan across an area of interest in the device and "search" for a signal, whereas conventional methods were probing at one single point and - depending whether the probe was placed to the right position - sometimes did not acquire any signal. The only information that need to be known for the measurement methods of this work is the frequency of the expected signal, then the SA can be set to this particular frequency and the signal extraction will be easy. This way, signals can be traced through the IC and entire areas of interest can be evaluated at the same time. Additionally, in the future, the layout of the device can be overlaid if necessary.

The MAMs determined the relative signal strength across an area. The MSMs are not capable of determining the signal strength, because this method only extracts the sign of signals. Nonetheless this measurement method showed a strong sensitivity, which enabled signal resolution of very faint signals that were not detected in the MAMs. A reason might be that the MSM tool only evaluates, whether there is a signal, which increases from the static value or not and provides a single boolean result. The MAM tool in opposition records signal levels for all positions, such that low signal levels might be misinterpreted as noise floor. Thus MSM might be very useful in the future for measurements of smaller devices, because - in case the signal levels decrease with the size of the device - the signal might still be detectable. As an example of signal detection of very faint signals, see section 9.4.1.

With the VSs it is possible to extract the signal-to-voltage correlations and the signal levels at the nominal device voltage with one single measurement. However it is not possible - as with the conventionally used time-domain measurements - to extract the sign of the signal or to tell something about the shape of the waveform. Further, especially for measurements of sub-micron devices, the probe placement plays an important role. Still, this tool enables fast (only a few seconds) and low-noise signal extraction.

For future tasks, the use of this tool can be expanded: the SA could extract signal levels of higher harmonics to determine the shape of the waveform. Further, for the scope to measure even unstable signals, e.g. from ring oscillators, it might be useful to build up a feed-back loop that constantly measures the frequency of the device and adjusts the SA to this frequency.

However, at least in the measurements of the devices with the nominal gate dimensions, it was found that strong signal levels occur, when both pins - gate and drain - are pulsed simultaneously (compare sections 9.2.3 and 9.3.3). This means that is is not only important to chose the right measurement method and the right probe place, but is is also vital to drive the device in the right mode, such that the strongest signal level can be detected.

Part VI.
Modeling

10. Active and passive signal contribution

In part III the LVP setup and measurement methods used for this work were described. This part explains active and passive signal generation (chapter 10) of the devices and shows, how the relevant effects that occur in a device, while a laser beam is focused on its active areas (see chapter 4), can be evaluated in the simulations of the reflectance from these structures (chapter 11 and 12).

10.1. Static part of the reflected light

In comparison to the modulated part of the reflected light the static part is orders of magnitude larger (ppm modulation compared to a few percent static). All the reflections within the depth of focus contribute to this overall static reflectance: each interface reflects a certain amount of light. The strength of the reflectance of one layer is depending on its index of refraction, the absorption coefficient and the thickness of the layer (whenever multi-reflections take place) - for details see 3.1. For most of the media that can be found in a CMOS structure, these properties are independent of the voltage that is applied: the silicide, oxides and metalizations. The properties of the silicon materials (bulk-Si, well, diffusion) depend on the voltage that is applied, but, even in the static condition, they depend on the effects, which were described in chapter 4 - i.e. the laser wavelength, the temperature, the electric field and the free carrier concentration. The corresponding properties of the materials will be described in detail in chapter 11.

10.2. Modulated part of the reflected light

The modulated part of the reflected light is the actual LVP signal. The modulation (**active signal contribution**) is caused by the voltage that is applied to the pins used - drain and / or gate of the MOSFET. Pulsing of the **drain** causes the **space charge region thickness** to change. The thickness of the space charge region influences the optical path length and thus plays an important role for interference effects. When the **gate** of a MOSFET is pulsed, again, the **space charge region thickness** underneath the gate is modulated in width and, in addition, the **inversion channel** develops, so the charge carrier density is modulated for voltages above the threshold voltage. The modulated part of the reflected light can be influenced by the static properties of the materials in the background, as well, by the interference (compare to section 3.2) of the reflected beams (**passive signal contribution**), e.g. when the two waves are phase-shifted by π. The following chapter will analyze various interfaces in detail and determine, whether such a phase shift is expected or not.

11. Separate modeling of the interfaces

As shown in chapter 3.2, a detailed reflectance calculation of a structure like a MOSFET is too complex to be done by hand. Parts of the structure will be simulated in the following chapter, to understand better, which parameters influence the reflectance. This chapter analyzes only the properties of single interfaces. The analysis will show the properties of silicon with different doping levels in detail, determine how the indices of refraction of the layers compare and conclude, whether there is a phase shift at the according interface or not. For layers that are actively contributing to the signal, the influence of the absorption coefficient, the refractive index and the thickness of the layers on the reflectance will be investigated in chapter 12.

As described in part IV, various devices have been investigated for this work. These devices consist of a fairly complex structure. For the modeling, simple layers with homogeneous properties and defined interfaces are needed. Doping profiles would be too complex, so the pn-junctions are treated as abrupt step junctions in which the doping profiles are replaced by average doping concentrations, the space charge regions are assumed to be fully depleted (depletion approximation) and the inversion channels are treated as layers of defined thicknesses and average inversion charge densities (for calculations of these parameters see section 7.4). These approximations will be used for the modeling and the simulations. Variations of the average doping levels will show the correlation with the signal in chapter 12.

Table 11.1 contains the index of refraction of the materials that can be found in a MOSFET. The according values for silicon have been extracted from the theory in chapter 4 for the two wavelengths used, 1319 nm (photon energy 0.94 eV) and 1064 nm (1.17 eV) (see information in brackets for the exact data points of extraction), and will be discussed in the following sections. For the materials cobalt silicide (CoSi) [1], silicon nitride (SiN) [2], silicon dioxide (SiO_2) [3] and aluminum (Al) [4] the sources can be found in the footnotes.

11.1. Effects in silicon

---
[1] reference: http://www.stormingmedia.us/38/3813/D381358.html.
[2] reference: http://www.ee.byu.edu/photonics/opticalconstants.phtml.
[3] reference: http://www.ee.byu.edu/photonics/opticalconstants.phtml.
[4] reference: http://www.ee.byu.edu/photonics/opticalconstants.phtml.

Table 11.1.: Indices of refraction for materials occurring in a FET (the short cut f.c. stands for "free carrier effects")

Wavelength [nm]	1319	1064
Silicon		
n_{Si} dispersion (intrinsic Si, 300 K)	3.501	3.568
$\frac{\Delta n}{\Delta T}$ thermo-optic $[K^{-1}]$ (295 K, 1300/1100 nm)	$1.94 \cdot 10^{-4}$	$2.15 \cdot 10^{-4}$
Δn electro-refraction ($10^5\ Vcm^{-1}$, 1320/1065 nm)	$\approx +0.25 \cdot 10^{-5}$	$\approx +0.8 \cdot 10^{-5}$
Δn Kerr (independent of λ, $10^5\ Vcm^{-1}$)	$\approx -10^{-6}$	
Δn f.c. ($3.2 \cdot 10^{17}$ electrons cm^{-3}, 1100 nm)	see below	$-1 \cdot 10^{-4}$
Δn f.c. ($4 \cdot 10^{19}$ electrons cm^{-3}, 1100 nm)	see below	$-2 \cdot 10^{-2}$
Δn f.c. ($3.2 \cdot 10^{17}$ electrons cm^{-3}, 1300 nm)	$-2 \cdot 10^{-4}$	see above
Δn f.c. ($4 \cdot 10^{19}$ electrons cm^{-3}, 1300 nm)	$-3 \cdot 10^{-2}$	see above
Δn f.c. ($5 \cdot 10^{17}$ holes cm^{-3}, 1100 nm)	see below	$-6 \cdot 10^{-4}$
Δn f.c. ($1 \cdot 10^{20}$ holes cm^{-3}, 1100 nm)	see below	$-4 \cdot 10^{-2}$
Δn f.c. ($5 \cdot 10^{17}$ holes cm^{-3}, 1300 nm)	$-9 \cdot 10^{-4}$	see above
Δn f.c. ($1 \cdot 10^{20}$ holes cm^{-3}, 1300 nm)	$-6 \cdot 10^{-2}$	see above
Other materials		
n_{CoSi}	3.2	
n_{SiN}	2	
n_{SiO_2}	1.45	
n_{Al}	1.23	
$n_{poly-Si} = n_{Si,intrinsic}$ (see above)	3.501	3.568

Thermo-optic coefficient Due to setup and measurement issues - see chapter 6 - the laser power was limited to maximum values: 7.4 mW (1064 nm laser wavelength) and 1.3 mW (1319 nm laser wavelength), to guarantee that the APD operates in linear mode, see chapter 6. The estimated resulting peak temperature variations in metal lines - as described in [BDPB04] [5] - are $4.07°C$ (1064 nm) and $0.72°C$ (1319 nm). Note that the calculations in the literature cited here were performed for 1300 nm laser wavelength, so the estimation for 1064 nm laser wavelength is not accurate. Note further that the peak temperature variations are calculated for metal only, but basically, during an LVP measurement, the laser is focused at the diffusions not at the metal lines, so the temperature increase will be even lower, because less light will be absorbed. A first level approximation of the thermo-optic coefficient, with the temperature variations from above, estimates $\Delta n = 8.75 \cdot 10^{-4}$ for 1064 nm and $1.40 \cdot 10^{-4}$ for 1319 nm laser wavelength. But even for a temperature increase of $10°C$ - for example in the inversion channel - the changes in the index of refraction would be $\Delta n = 1.94 \cdot 10^{-3}$ and $2.15 \cdot 10^{-3}$ for 1319 nm and 1064 nm respectively. In addition, Goldstein et al. ([GSG93]) presented data, which showed that changes in the index of refraction due to temperature variations are only measurable, if current flows (they measured bipolar transistors!), otherwise free carrier effects dominate the signal. And - with a phase sensitive detection scheme - they were able to measure the electro-optic effect, but found that the signal levels are still lower than the signal levels produces by the free carrier effects. For a bipolar structure measured with a ML laser they found $\Delta T = 6.4$ K, which lies within the estimated limits.

Kerr effect and electro-refraction With about $\Delta n = -10^{-6}$ the **Kerr effect** plays a minor role among the field effects that occur in silicon. The evaluation of **electro-refraction** predicts a maximum change in the index of about $\Delta n = 0.8 \cdot 10^{-5}$ for 1065 nm laser wavelength and electric field strengths of $10^5 \ \frac{V}{cm}$. The peak values, which can be extracted from figure number 4.6, are $+\Delta n = 1.3 \cdot 10^{-5}$ and $-\Delta n = 1.7 \cdot 10^{-5}$ (for $10^5 \ \frac{V}{cm}$). Even with the prediction that the effect is 2x stronger for $E_{opt} \parallel E_{appl}$ than for $E_{opt} \perp E_{appl}$ ([SB87]), this leads to $+\Delta n = 2.6 \cdot 10^{-5}$ and $-\Delta n = 3.4 \cdot 10^{-5}$ (for $10^5 \ \frac{V}{cm}$).

Free carrier refraction For charge carrier densities higher than $6 \cdot 10^{17} \ cm^{-3}$ for holes and $1-2 \cdot 10^{18} \ cm^{-3}$ for electrons (well doping concentration: $2 \cdot 10^{17} \ cm^{-3}$!), the change in the refractive index due to **free carrier effects** is higher than for the effects mentioned above: $\Delta n > 1 \cdot 10^{-3}$. This means that the free carrier effects **play the major role in the change of the refractive index** and hence the influence on the reflectance.

With this information about the indices of refraction, a first order evaluation of the resulting phase shift of the reflected wave ($0°$ or $180°$ phase shift between incident wave and reflected wave) at the according interfaces for the electric field perpendicular or

---

[5] For one particular structure - a metal line - the peak temperature variation induced by laser heating was calculated to be $\Delta T_{max} = \frac{0.55°C}{mW}$.

parallel to the incident plane can be achieved. Coming from the backside of the FETs, the light traverses the interfaces listed in table 11.2; the phase shifts are predicted by using the Fresnel equations (see equation 3.5 and 3.2), which evaluate the relation of the indices of refraction for the medium of incidence and the medium of transmission. The material layers differ slightly for the NFET and the PFET due to the opposite well doping type:

PFET

- p^--type substrate – SCR: this interface can only be found in PFETs, because there is a pn-junction between the substrate (p) and the well (n).
- SCR – n-type well: also only in PFETs (see above).

NFET

- p^--type substrate – p-type well: only in NFETs: there is no SCR between the substrate and the well, but a discontinuity.

PFET and NFET

In the drain / source area:

- well – SCR: for the first order approximation of the index of refraction, the doping type is irrelevant, because the index will decrease from its intrinsic value for both types (see negative sign in equation number 4.5)
- SCR – diffusion: again, the doping type is not relevant
- diffusion – silicide
- silicide – nitride
- silicide – interconnect metal: not shown in figure number 7.1; these are the contacts from the diffusion to the metal lines connecting the FETs with the outer circuitry

In the gate area:

- well – SCR: the doping type is irrelevant (see above)
- SCR – inversion channel (if there is a channel)
- channel – oxide: the interface to the gate dielectric, or
- SCR – oxide (without channel)
- oxide – poly-Si gate

- poly-Si gate – silicide
- silicide – nitride
- silicide – interconnect metal: connecting the gate with the outer circuitry

From table 11.2 the phase shifts at the interfaces occurring in a FET can be determined for the two polarization states (compare to section 3.1). Assuming that the bulk of the laser light is either absorbed or reflected by the metal and the silicide, the following assumptions can be made. In the π-case, for the NFET and the PFET, the phase shifts at 4 interfaces in the gate area (interface to the well, the channel, the oxide and the silicide) but also in the drain area (interface to the well, the diffusion, the silicide and the aluminum). So the part of the reflection caused by the electric field component perpendicular to the incident plane indicates no phase shift between the PFET and the NFET nor does it predict a phase shift between the drain and the gate area. In the σ-case, the according properties of the materials suggest that the phase of the reflected wave will shift by 180° at the interface to the SCR, the metal and the poly-Si gate. If there were no additional interference effects, this would mean that the LVP signals of gate and drain were in phase (phase shift at the SCR for both, gate and drain and in the gate at the poly-Si, in the drain at the metal) and that the LVP signals of the PFET and NFET differ, because, only at the PFET there is an additional SCR between p-substrate and n-well, where the phase shifts again.

In fact, the interference effects of the active signal sources alone will most likely already differ for gate and drain signals and thus a prediction of the signal behavior of the overall reflectance for both is hardly possible. (Interference can take place for layers that are thicker than 76 nm (for 1064 nm wavelength) or 94 nm (for 1319 nm), as can be derived from equation number 3.12, the difference in path length.)

Both parts of the reflections (π-case and σ-case) are detected for the amplitude information, so the overall reflectance is influenced by the resulting phase shift between the two waves in addition. A phase shift could only occur, if the difference in path length varies for the two polarizations states - which is not possible - or if the properties vary for the two states, which was only predicted for electro-refraction (see section 4.4.2), but this effect is assumed to be negligible in comparison to the free carrier effects.

Table 11.2.: Interfaces occurring in a FET, the according relation of the indices of refraction and the resulting phase of the reflected wave at the interface (for metal: 180° phase shift due to the properties of the metal not due to the index of refraction.)

Interfaces medium 1 - medium 2	Relation of the indices of refraction medium 1 - medium 2 [-]	Phase between incident and reflected wave π-case [°]	Phase between incident and reflected wave σ-case [°]
PFET			
p^--type substrate – SCR	$n_{sub} < n_{SCR}$	0	180
SCR – n-type well	$n_{SCR} > n_{well}$	180	0
NFET			
p^--type substrate – p-type well	$n_{sub} > n_{well}$	180	0
PFET and NFET			
well – SCR (underneath the gate or the drain)	$n_{well} < n_{SCR}$	0	180
SCR – drain diffusion / inversion channel	$n_{SCR} > n_{diff/chan}$	180	0
drain diffusion – silicide (CoSi)	$n_{diff} > n_{CoSi}$	180	0
silicide (CoSi) – nitride (SiN)	$n_{CoSi} > n_{SiN}$	180	0
silicide (CoSi) – interconnect metal (Al)	$n_{CoSi} > n_{Al}$	180	180
channel – oxide	$n_{chan} > n_{SiO_2}$	180	0
oxide – poly-Si gate	$n_{SiO_2} < n_{poly}$	0	180
poly-Si gate – silicide (CoSi)	$n_{poly} > n_{CoSi}$	180	0

12. Simulations with the matrix formalism

First, this chapter will explain, how the modeling of the structures that were simulated - NFET and PFET, drains and gates - was done. Then it will be discussed, how the matrix formalism was used to perform the simulations of the reflectance. The simulations were done for both wavelengths, 1319 nm and 1064 nm. The results will be analyzed in detail and the main influencing parameters will be extracted.

12.1. Using the matrix formalism for reflectance simulations of the active areas of a FET

As described in section 3.2.2, for the calculation of the overall reflectance of a structure, it is necessary to define simple layers with interfaces and their according properties. This chapter will analyze the active part of the signal, which is caused by the electrical activity of the device and not by the background signals (compare to passive signal contribution in chapter 10 and 11).

12.1.1. Modeling of the drains of the FETs

The only active signal source in the drain area can be reduced to the modulation of the SCR width due to the drain voltage. This assumption leads to a structure with only two interfaces: the well-to-SCR interface and the SCR-to-diffusion interface. Since in chapter 11.1 the main effect that influences the properties of silicon was determined to be the free carrier effect, the properties of these interfaces can be calculated from equations number 4.5 and 4.4 and with the SCR thickness. The situation is shown in the schematic in figure 12.1.

12.1.2. Modeling of the gates of the FETs

For the gates of the FETs, there are two different situations that need to be taken into account:

- if the gate voltage is lower than the threshold voltage, the only active signal source is the SCR below the gate that is modulated in width. The according interfaces then are: the well-to-SCR interface and the SCR-to-**oxide** interface. See figure number 12.2.

- if the gate voltage is higher than the threshold voltage, the active signal source is the modulation of the inversion charge carrier density in the channel, since the SCR thickness stays constant for voltages above the threshold. The according

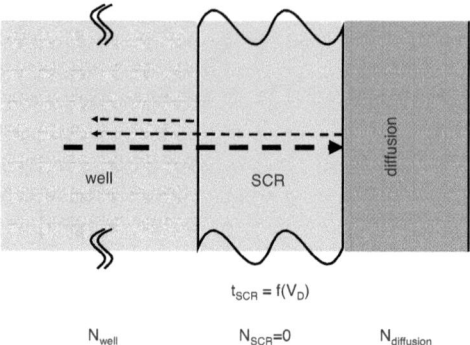

Figure 12.1.: Schematic diagram of the simplified diffusion structure for the simulation

interfaces then are: the well-to-SCR interface and the SCR-to-**channel** interface. See figure number 12.3.

Both models assume that there is no electric field component at the right side of the last layer (see boundary condition of the matrix formalism 3.2.2).

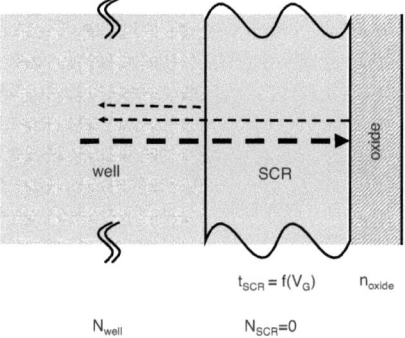

Figure 12.2.: Schematic diagram of the simplified gate structure for the simulation in the sub-threshold regime

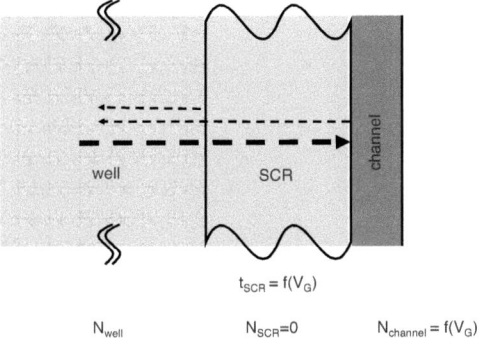

Figure 12.3.: Schematic diagram of the simplified gate structure for the simulation (voltages above threshold voltage)

The optical parameters of all layers can be calculated from the equations number 4.5 and 4.4, which determine the indices of refraction and the absorption coefficient (and thus the extinction coefficient) of the layers due to free carriers. Note that for the two wavelengths, dispersion needs to be taken into account, too. With the refractive index and the extinction coefficient, the reflection and transmission coefficients of the interfaces are calculable and hence the transfer matrices can be evaluated. For the propagation matrix, the phase factor needs to be calculated, which includes the interference effects due to the variations in the thickness of the layers: the SCR thicknesses. The equations that were used to determine these values for the gate and the drain are explained in section 7.4.

12.1.3. Modeling of mobility for various carrier concentrations

For the calculation of the absorption coefficient from equation number 4.4, the mobility of the charge carriers is needed. Since the mobility is depending on the carrier density and the material - i.e. the impurity (or doping) concentration and type of silicon, but also the free carrier concentration - the according densities need to be determined. So when the doping concentration and the inversion channel charge carrier density are known, the mobility can be evaluated. In the literature, various sources for calculations of the mobility from those properties can be found. For the calculations in this work, the data from Wolf, which can be found in [Wol69], was used (formulas see appendix). Figure number 12.4 shows the mobility versus charge carrier density for holes and electrons in p- and n-silicon. The degeneracy of the semiconductor (very high charge carrier densities, starting from around $2 \cdot 10^{19}$ cm^{-3}) is taken care of by limiting the mobility at the lower end: 140 $\frac{cm^2}{Vs}$ for electrons and 100 $\frac{cm^2}{Vs}$ for holes.

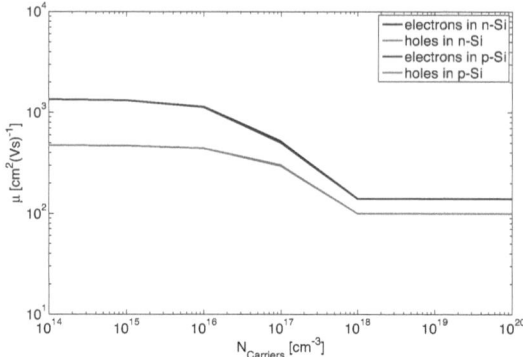

Figure 12.4.: Mobility versus charge carrier density for holes and electrons in n- and p-silicon after Wolf [Wol69]

12.1.4. Calculation of the refractive index and the absorption coefficient

With the calculated charge carrier densities and the mobilities for those values, the changes in the index of refraction and the absorption coefficient for holes and electrons for the two wavelengths used (1319 nm and 1064 nm) can be calculated from equations number 4.5 and 4.4. Figures number 12.5 and 12.6 show the calculations and the extracted values from [SB87] for 1300 nm wavelength for comparison. As discussed section 4.5, the mentioned measurements do not agree exactly with the theoretical calculations, but still, for the modeling, the extraction of the refractive index and the absorption coefficient from the equations was chosen. This enables simulations for both wavelengths used here, whereas the measurements only predict the data for 1300 nm and 1550 nm laser wavelength.

For one example (the varactor) of the calculation development see the diagram in figure 12.7.

12.2. Simulation results: n^+p-diode, reverse bias (NFET, drain)

Figure number 12.8 and 12.9 show the result of the simulated reflectances at the interfaces of the diode in reverse bias for the NFET versus the junction voltage for the two wavelengths. The average of the diffusion doping concentration was set to $2 \cdot 10^{20}$ cm^{-3}, the well doping concentration was set to $2 \cdot 10^{17}$ cm^{-3} - the extracted values from the

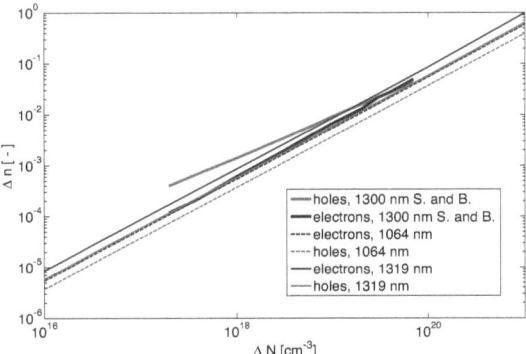

Figure 12.5.: Changes in the index of refraction due to free carriers versus charge carrier density for holes (red) and electrons (blue) and 1319 nm (solid lines) and 1064 nm (dashed lines) wavelength. The values extracted from Soref and Bennett ([SB87]) for 1300 nm are shown as well.

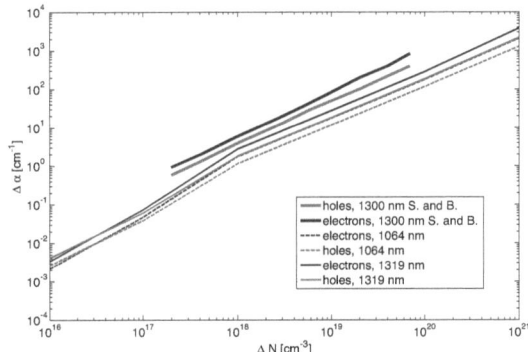

Figure 12.6.: Changes in the absorption coefficient due to free carriers versus charge carrier density for holes (red) and electrons (blue) and 1319 nm (solid lines) and 1064 nm (dashed lines) wavelength. The values extracted from Soref and Bennett ([SB87]) for 1300 nm are shown as well.

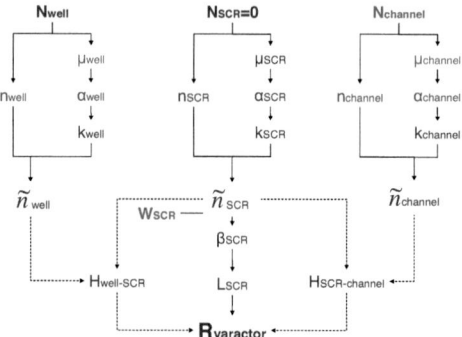

Figure 12.7.: Development of the calculations for the simulations: from charge carrier density to the overall reflectance, calculation with the matrix formalism. The blue parameters indicate the dependency on free carrier effects; red parameters label the electrical data, which need to be evaluated first. The SCR calculations are performed with the intrinsic data set (no free carrier effects).

electrical simulation data (compare to section 7.3, referred to as the "original values" in the following). The reflectance of the 1319 nm simulation increases up to 0.6 V, but then it decreases again. The maximum amplitude of the simulation is with 0.22 ppm over a factor of 500 lower than the measured signal (the drain signal of the NF-L-120 was linear, positive and had an amplitude of around 125 ppm, see section 9.1.1). The result of the 1064 nm laser wavelength simulation is linear with a negative sign, which is comparable to the according measurement (see section 9.1.1), but the amplitude is again lower than expected - this time it is a factor of around 60.

Since the simulations with the original values of the average doping concentrations were not reproducing the results of the measurements, in the following, the well and the diffusion concentrations were varied to investigate the dependency of the reflectance on the doping concentrations. First, the well doping concentration was set to $2 \cdot 10^{17}$ cm^{-3} and the diffusion doping concentration was varied from $5 \cdot 10^{19}$ to $1 \cdot 10^{22}$ cm^{-3} for the 1319 nm laser and from $5 \cdot 10^{19}$ to $2 \cdot 10^{21}$ cm^{-3} for the 1064 nm laser (values as labeled in the graphs). The results are shown in figure 12.10 and 12.11. For easier comparison, the simulations with the original values are shown in the graphs, too.

The simulations for the 1064 nm laser are linear in shape (as the original signal) and increasing in amplitude (absolutely, as the result is negative) for increased diffusion dop-

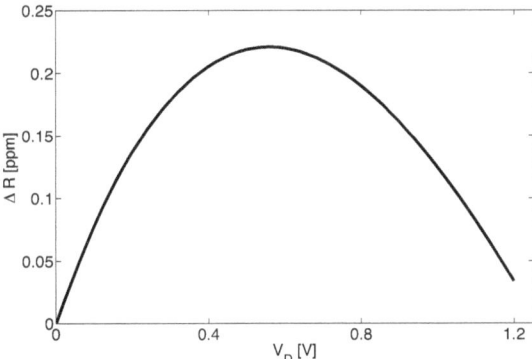

Figure 12.8.: Simulation for 1319 nm laser wavelength - reflectance at the interfaces of the reverse biased diode (NFET, drain), well and diffusion doping concentration set to the values extracted from the electrical simulations: $2 \cdot 10^{17}$ and $2 \cdot 10^{20}$ cm^{-3}

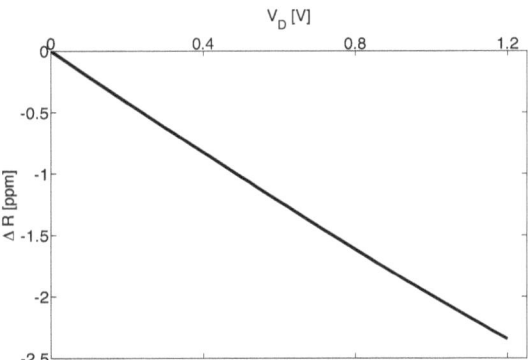

Figure 12.9.: Simulation for 1064 nm laser wavelength - reflectance at the interfaces of the reverse biased diode (NFET, drain), well and diffusion doping concentration set to the values extracted from the electrical simulations: $2 \cdot 10^{17}$ and $2 \cdot 10^{20}$ cm^{-3}

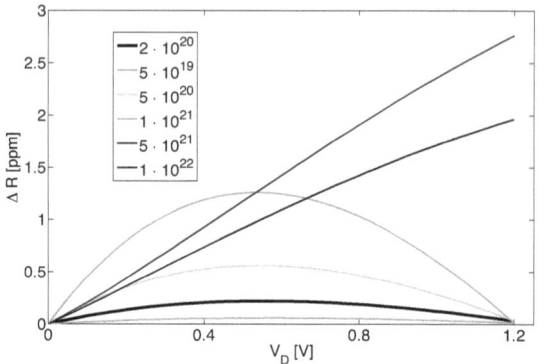

Figure 12.10.: Simulation for 1319 nm laser wavelength - reflectance at the interfaces of the reverse biased diode (NFET, drain), well doping concentration set to $2 \cdot 10^{17}$ cm^{-3}, diffusion doping concentration varied

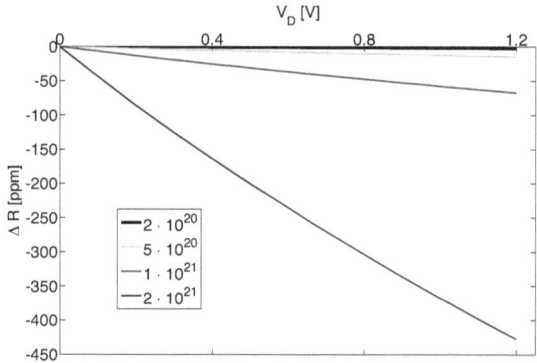

Figure 12.11.: Simulation for 1064 nm laser wavelength - reflectance at the interfaces of the reverse biased diode (NFET, drain), well doping concentration set to $2 \cdot 10^{17}$ cm^{-3}, diffusion doping concentration varied

ing concentrations. The maximum amplitude of the simulation reaches -472 ppm for a diffusion doping concentration of $2 \cdot 10^{21}$ cm^{-3}. The results of the simulations for the

1319 nm laser wavelength are of the same shape as the simulation of the original values, but with amplitudes that are increasing with higher diffusion doping concentration. For diffusion doping concentrations higher than $5 \cdot 10^{21}$ cm^{-3} the shape of the simulation is almost linear. With 2.76 ppm, even the strongest amplitude of the simulations (for $5 \cdot 10^{21}$ cm^{-3}) is still very weak compared to the measurements. For diffusion doping concentrations above $5 \cdot 10^{21}$ cm^{-3} the amplitude of the simulation even decreases again. To investigate, how the strongest amplitude of the simulation can be achieved, the diffusion doping concentration was varied even further (see figure 12.12).

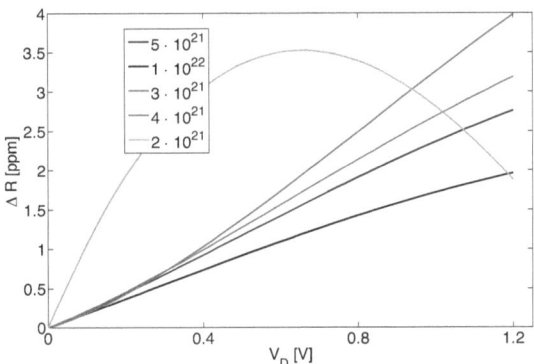

Figure 12.12.: Simulation for 1319 nm laser wavelength - reflectance at the interfaces of the reverse biased diode (NFET, drain), well doping concentration set to $2 \cdot 10^{17}$ cm^{-3}, diffusion doping concentration varied at a very high level

The simulations for diffusion doping concentrations above $3 \cdot 10^{21}$ cm^{-3} are linear, but decrease further for an increased doping concentration. For a doping concentration of $2 \cdot 10^{21}$ cm^{-3}, the slope of the simulation is very steep at voltages lower than 0.4 V, but it decreases as the voltage increases. In opposition to the result of the simulations for $1 \cdot 10^{21}$ cm^{-3} the simulation does not reach 0 ppm at the nominal voltage, but reaches around 2 ppm at 1.2 V.

Since the variations of the diffusion doping concentration already showed a strong influence on the simulation results, in the following, the diffusion doping concentration was held constant and the well doping concentration was varied to in investigate the influence on the simulation result. The values of the diffusion doping concentrations were set to $2 \cdot 10^{21}$ cm^{-3} in order to get the closest match with the measurements. The results are shown in graph 12.13 and 12.14.

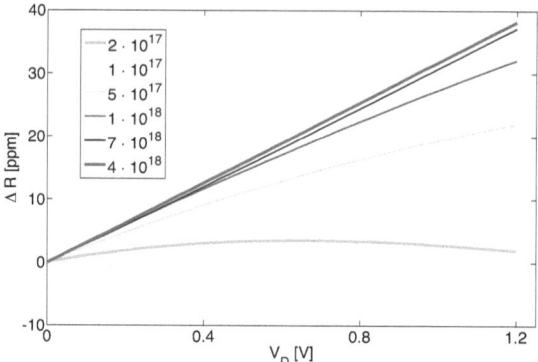

Figure 12.13.: Simulation for 1319 nm laser wavelength - reflectance at the interfaces of the reverse biased diode (NFET, drain), diffusion doping concentration set to $2 \cdot 10^{21}$ cm^{-3}, well doping concentration varied

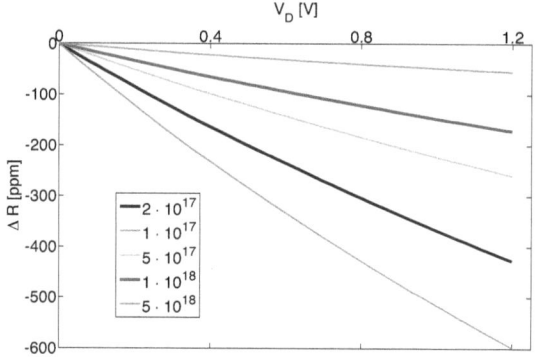

Figure 12.14.: Simulation for 1064 nm laser wavelength - reflectance at the interfaces of the reverse biased diode (NFET, drain), diffusion doping concentration set to $2 \cdot 10^{21}$ cm^{-3}, well doping concentration varied

The simulation results of the 1319 nm wavelength vary a lot in shape, sign and amplitude. From $2 \cdot 10^{17}$ to $4 \cdot 10^{18}$ cm^{-3} the amplitude increases drastically. For well doping concentrations higher than $4 \cdot 10^{18}$ cm^{-3} (e.g. see result for $7 \cdot 10^{18}$ cm^{-3}), the amplitude of the simulation decreases again. For a well doping concentration of $1 \cdot 10^{17}$ cm^{-3} the simulation even predicts negative values. All results of the simulations below and above the original well doping concentration have quite linear signal-to-voltage correlations. The closest match with the measurements results from the simulation with $4 \cdot 10^{18}$ cm^{-3} well doping concentration and $2 \cdot 10^{21}$ cm^{-3} diffusion doping concentration (labeled red in the graph): the signal-to-voltage correlation is linear and the sign is positive, but with an amplitude of 38 ppm, it is still a factor of 3.3 lower than the result of the measurement.

The 1064 nm wavelength simulation results predict a decreasing (negative) amplitude for an increased well doping concentration. The signal-to-voltage correlation stays linear for all values. For a well doping concentration of $1 \cdot 10^{18}$ cm^{-3} the amplitude of the simulation reaches -171 ppm, which compares well with the measurements: the measured signal was linear, had a negative sign and an amplitude of around 150 ppm.

12.2.1. Influencing parameters

The sign, the shape and the amplitude of the simulation is influenced by different parameters. If the doping concentrations of the layers are varied, as in the simulations shown above, the absorption due to free carriers $\Delta\alpha$, the index of refraction Δn and the thickness of the space charge region t_{SCR}, but also the built-in potential and the mobility of the charge carriers are influenced, since all of them depend on the carrier density. This section investigates, how far and in which way, these parameters affect the simulation. The investigation is based on two simulations with the same diffusion doping concentration ($2 \cdot 10^{21}$ cm^{-3}) and different well doping concentrations: a) $2 \cdot 10^{17}$ and b) $4 \cdot 10^{18}$ see black and red curve in figure 12.15. Starting at curve a), for a well doping concentration of $2 \cdot 10^{17}$ cm^{-3}, the change in the reflectance due to the above mentioned parameters is extracted. The entire reflectance calculation is done with $2 \cdot 10^{17}$ cm^{-3}, only the parameter of interest is calculated with the higher doping concentration $4 \cdot 10^{18}$ cm^{-3}. For example, if the space charge region thickness is calculated with $4 \cdot 10^{18}$ cm^{-3}, the simulated curve shows the reflectance change due to the SCR thickness only. The same is done for the index of refraction, the built-in potential and the mobility. The direct influence on the absorption coefficient can not be determined in this way, because the calculation contains the changes in the index of refraction and the mobility as well, so the effects mix (equation 4.4 shows that the absorption coefficient is calculated with the index of refraction and the mobility, both depend on the charge carrier density). But the influence can be extracted indirectly, as will be shown below. The results of the simulations for 1319 nm laser wavelength can be found in figure number 12.15. Note that the simulations for 1064 nm follow the same pattern (no diagram).

The built-in potential and the mobility both influence the amplitude of the simulation marginally and thus are not shown in the graph. The thickness of the SCR is mostly influencing the shape of the simulation (from parabolic to linear signal-to-voltage corre-

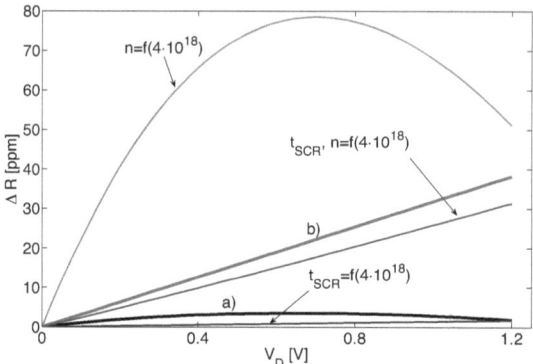

Figure 12.15.: Simulation for 1319 nm laser wavelength - reflectance at the interfaces of the reverse biased diode (NFET, drain), diffusion doping concentration set to $2 \cdot 10^{21}$ cm^{-3}, well doping concentration varied between $2 \cdot 10^{17}$ (a) black curve) and $4 \cdot 10^{18}$ cm^{-3} (b) red curve) and the influence of the well doping concentration on the amplitude and shape of the simulation (via dependencies of the SCR, index of refraction)

lation). The index of refraction has an effect on the amplitude, but the signal-to-voltage correlation stays almost the same. The violet curve shows the combined effect of the SCR and the index of refraction - it matches the goal simulation b) closely. Since the built-in potential and the mobility only play a minor role, from the latter simulation, the influence of the absorption coefficient can be extracted inductively: it affects the amplitude of the simulation, but not as strong as the index of refraction.

The same evaluation has been done for two different diffusion doping concentrations and a fixed well doping concentration, but the results predict that the thickness of the SCR has only a minor influence on the simulation, because a variation of the higher doping concentration has only a small effect on the calculation of the SCR thickness. The index of refraction causes the main influence on the simulation, whereas the absorption coefficient again plays the minor role. The results are not shown here, due to the small difference between the curves.

12.2.2. Summary

The simulation results are strongly depending on the well and diffusion doping concentrations. The results vary in amplitude, sign and shape. Graphs number 12.16 and 12.17 show the simulations, which match the measurements the best. The diffusion concentration was set to $2 \cdot 10^{21}$ cm^{-3} for both wavelengths and the well doping concentration was set to $4 \cdot 10^{18}$ cm^{-3} for the 1319 nm and to $1 \cdot 10^{18}$ cm^{-3} for the 1064 nm wavelength simulation. Both doping concentrations - for well and diffusion - are well above the original values.

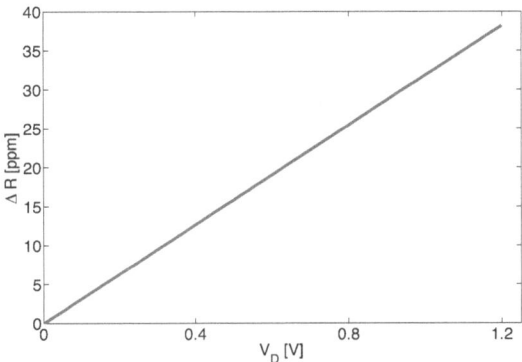

Figure 12.16.: Simulation for 1319 nm laser wavelength - reflectance at the interfaces of the reverse biased diode (NFET, drain), diffusion doping concentration set to $2 \cdot 10^{21}$ cm^{-3}, well doping concentration set to $4 \cdot 10^{18}$ cm^{-3}, matches the measurement in sign and shape, but has a lower amplitude

The main influencing parameters are the SCR thickness - changes the signal-to-voltage correlation - and the index of refraction, which affects the amplitude of the simulation. The absorption coefficient also influences the amplitude of the simulation, but not as strong as the index of refraction. The built-in potential and the mobility only play a minor role.

12.3. Simulation results: p^+n-diode, reverse bias (PFET, drain)

The results of the simulated reflectances at the interfaces of the diode in reverse bias for the PFET versus the junction voltage for the two wavelengths can be found in figure number 12.18 and 12.19. As in section 12.2, the averages of the diffusion and well doping concentrations were set to the original values ($2 \cdot 10^{20}$ and $2 \cdot 10^{17}$ cm^{-3}). The results for both wavelengths are very similar to those of the NFET drain simulations (see figure 12.8 and 12.9). The amplitude of the 1064 nm wavelength simulation is slightly higher here than in the simulation of the NFET.

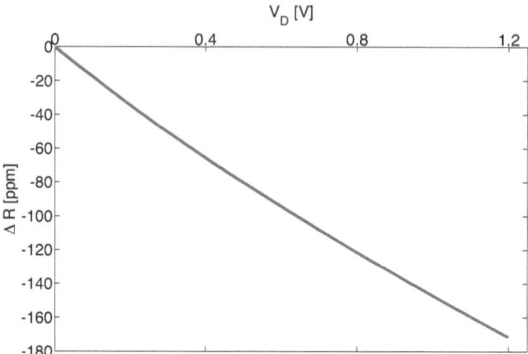

Figure 12.17.: Simulation for 1064 nm laser wavelength - reflectance at the interfaces of the reverse biased diode (NFET, drain), diffusion doping concentration set to $2 \cdot 10^{21}$ cm^{-3}, well doping concentration set to $1 \cdot 10^{18}$ cm^{-3}, matches the measurement in sign, shape and amplitude

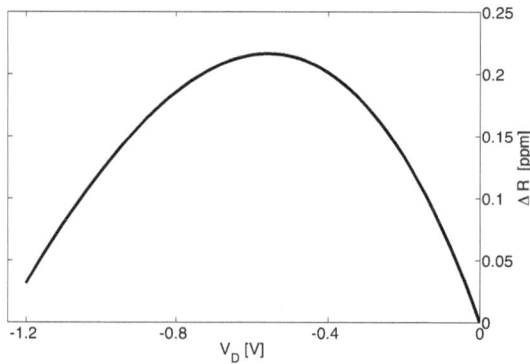

Figure 12.18.: Simulation for 1319 nm laser wavelength - reflectance at the interfaces of the reverse biased diode (PFET, drain), well and diffusion doping concentration set to the values extracted from the electrical simulations: $2 \cdot 10^{17}$ and $2 \cdot 10^{20}$ cm^{-3}

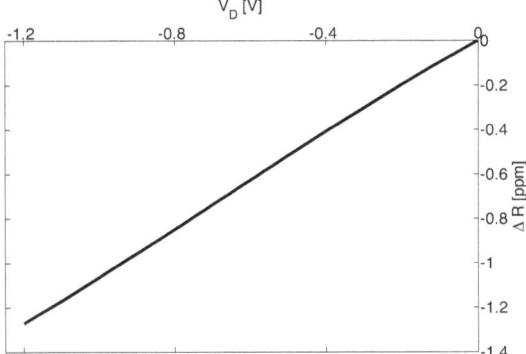

Figure 12.19.: Simulation for 1064 nm laser wavelength - reflectance at the interfaces of the reverse biased diode (PFET, drain), well and diffusion doping concentration set to the values extracted from the electrical simulations: $2 \cdot 10^{17}$ and $2 \cdot 10^{20}$ cm^{-3}

The measured drain signals of the PF-L-120 were linear and negative for both wavelengths and had amplitudes of around 75 ppm (1319 nm) and 150 ppm (1064 nm), see section 9.1.1. Since the simulations above, with the original values of the average doping concentrations, were not reproducing the results of the measurements - as for the NFET -, the well and the diffusion doping concentrations were varied to investigate the dependency on the reflectance in the following. The well doping concentration was set to $2 \cdot 10^{17}$ cm^{-3} and the diffusion doping concentration was varied from $1 \cdot 10^{20}$ to $1 \cdot 10^{22}$ cm^{-3} for the 1319 nm laser and from $1 \cdot 10^{20}$ to $2 \cdot 10^{21}$ cm^{-3} for the 1064 nm laser (values as labeled in the graphs). The results can be found in figure 12.20 and 12.21. For better comparison, the simulations with the original values are shown in the graphs, too.

The results of the 1319 nm simulations are very similar to those of the NFET with 1319 nm: the shapes and the signs are the same for same diffusion doping concentrations, but the amplitudes simulated here are a factor of two larger. The strongest amplitude was found for a diffusion doping concentration of $4 \cdot 10^{21}$ cm^{-3}, the signal-to-voltage correlation is almost linear, so this diffusion doping concentration was taken for the following investigations of the influence of the well doping concentration. The simulations for the 1064 nm laser are also similar to those of the NFET; the signal-to-voltage correlation is linear and the (negative) amplitude increases with increased diffusion doping concentration, but the amplitudes of the simulations here are around half as high for same concentrations. The simulation with $2 \cdot 10^{21}$ cm^{-3} already matches the signal amplitude of the measurements, which was about 150 ppm.

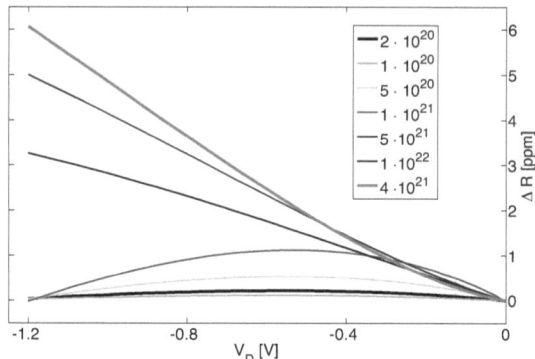

Figure 12.20.: Simulation for 1319 nm laser wavelength - reflectance at the interfaces of the reverse biased diode (PFET, drain), well doping concentration set to $2 \cdot 10^{17}\ cm^{-3}$, diffusion doping concentration varied

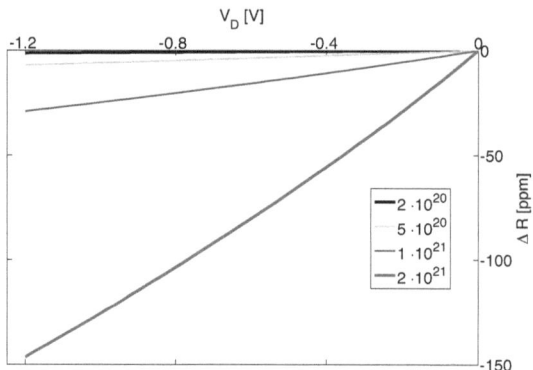

Figure 12.21.: Simulation for 1064 nm laser wavelength - reflectance at the interfaces of the reverse biased diode (PFET, drain), well doping concentration set to $2 \cdot 10^{17}\ cm^{-3}$, diffusion doping concentration varied

As in the simulations for the NFET, the variations of the diffusion doping concentration showed a strong influence on the simulations, so in the following, the diffusion doping concentration was held constant and the well doping concentration was varied in order to in investigate the effect on the simulation further. The values of the diffusion doping concentrations were set to $4 \cdot 10^{21}$ cm^{-3} for the 1319 nm simulations and $2 \cdot 10^{21}$ cm^{-3} for the 1064 nm simulations to get the closest match with the measurements. Graphs number 12.22 and 12.23 contain the results.

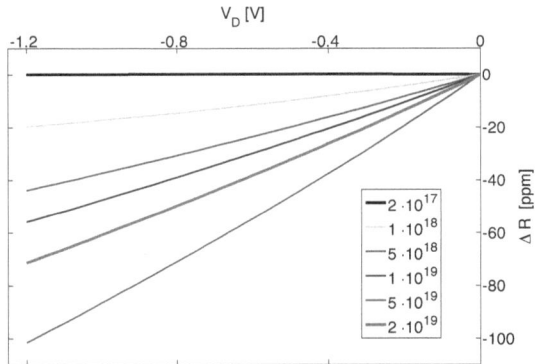

Figure 12.22.: Simulation for 1319 nm laser wavelength - reflectance at the interfaces of the reverse biased diode (PFET, drain), diffusion doping concentration set to $4 \cdot 10^{21}$ cm^{-3}, well doping concentration varied

The (negative) amplitudes of the 1319 nm wavelength simulations increase for an increased well doping concentration. The signal-to-voltage correlation stays linear for all values. For a well doping concentration of $2 \cdot 10^{19}$ cm^{-3} the amplitude reaches -71 ppm, which compares well with the measurements: the measured signal was linear, had a negative sign and an amplitude of around 75 ppm.

The (negative) amplitudes of the 1064 nm wavelength simulations increase for a decreased well doping concentration. The signal-to-voltage correlation stays almost linear for all values. The simulation with the well doping concentration set to the originally extracted $2 \cdot 10^{17}$ cm^{-3} still matches the measurements the best: the result is almost linear, negative and with an amplitude of -147 ppm very close to the amplitude of the measured signal, which was about 150 ppm.

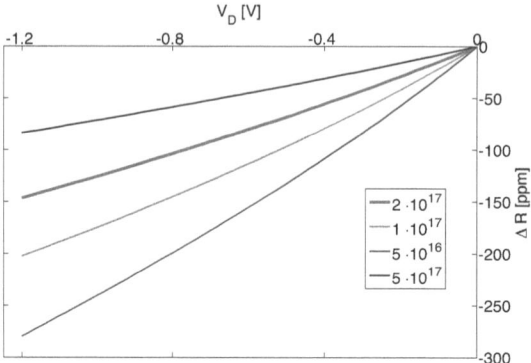

Figure 12.23.: Simulation for 1064 nm laser wavelength - reflectance at the interfaces of the reverse biased diode (PFET, drain), diffusion doping concentration set to $2 \cdot 10^{21}$ cm^{-3}, well doping concentration varied

12.3.1. Influencing parameters

As in section 12.2.1, the effects of the absorption, the index of refraction, thickness of the SCR, built-in potential and the mobility on the reflectance are investigated - again, as an example, with the 1319 nm laser wavelength. All simulations were performed with the same diffusion doping concentration ($4 \cdot 10^{21}$ cm^{-3}) and different well doping concentrations: a) $5 \cdot 10^{17}$ and b) $2 \cdot 10^{19}$ cm^{-3}, see black and red curve in figure 12.24. Starting at curve a), for a well doping concentration of $5 \cdot 10^{17}$ cm^{-3}, the change in the reflectance due to these parameters is extracted. The entire reflectance calculation is done with $5 \cdot 10^{17}$ cm^{-3}, only the parameter of interest is calculated with the higher doping concentration $2 \cdot 10^{19}$ cm^{-3}. The results of the simulations are shown in figure number 12.24. Again, the simulations for 1064 nm follow the same pattern (see section 12.2.1).

As found in the simulations for the NFET, the built-in potential and the mobility affect the amplitude of the simulation only marginally. The influence of the SCR thickness and the index of refraction (and extracted from those the absorption coefficient) also follow the same pattern as the simulations of the NFET: the thickness of the SCR mainly has an effect on the shape of the simulation (from parabolic to linear signal-to-voltage correlation) and the index of refraction affects the amplitude, but the signal-to-voltage correlation stays almost the same. The absorption coefficient has an effect on the amplitude of the simulation, but not as strong as the index of refraction.

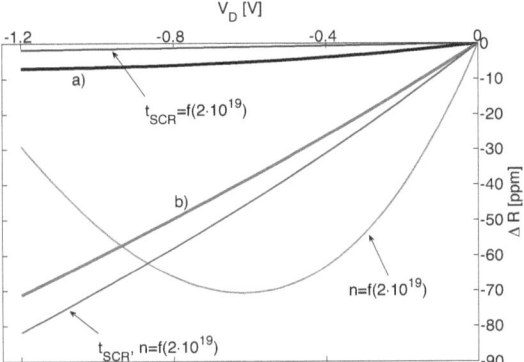

Figure 12.24.: Simulation for 1319 nm laser wavelength - reflectance at the interfaces of the reverse biased diode (PFET, drain), diffusion doping concentration set to $4 \cdot 10^{21}$ cm^{-3}, well doping concentration varied between $5 \cdot 10^{17}$ (black curve) and $2 \cdot 10^{19}$ cm^{-3} (red curve) and the influence of the well doping concentration on the amplitude and shape of the simulation (via dependencies of the SCR, index of refraction)

12.3.2. Summary

As the simulations of the NFET drain, the results of the PFET drain simulations are strongly depending on the well and diffusion doping concentrations. The results vary in amplitude, sign and shape. Graphs number 12.25 and 12.26 show the simulations, which match the measurements the best. For 1319 nm wavelength, the diffusion concentration was set to $4 \cdot 10^{21}$ and the well to $2 \cdot 10^{19}$ cm^{-3} and for 1064 nm wavelength to $2 \cdot 10^{21}$ and $2 \cdot 10^{17}$ cm^{-3}. Except the well doping concentration of the 1064 nm simulation, all doping concentrations - for well and diffusion - are well above the original values.

As for the NFET, the main influencing parameters that affect the simulation were extracted. The results are similar to those of the NFET. The SCR thickness mainly changes the signal-to-voltage correlation, the index of refraction and the absorption coefficient both influence the amplitude of the simulation, whereas the effect of the latter is smaller. The built-in potential and the mobility only play a minor role.

12.4. Simulation results: NFET, gate

Section 12.1.2 described how the simulations of the reflectance at the interfaces of the varactor will proceed. First, sub-threshold simulations will be shown, followed by the

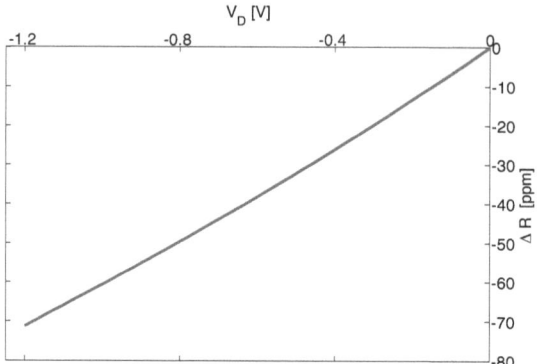

Figure 12.25.: Simulation for 1319 nm laser wavelength - reflectance at the interfaces of the reverse biased diode (PFET, drain), diffusion doping concentration set to $4 \cdot 10^{21}\ cm^{-3}$, well doping concentration set to $2 \cdot 10^{19}\ cm^{-3}$, matches the measurement in sign, shape and amplitude

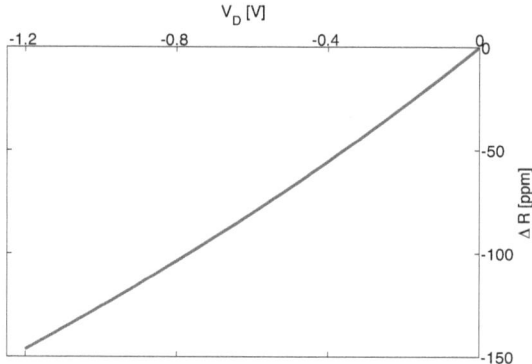

Figure 12.26.: Simulation for 1064 nm laser wavelength - reflectance at the interfaces of the reverse biased diode (PFET, drain), diffusion doping concentration set to $2 \cdot 10^{21}\ cm^{-3}$, well doping concentration set to $2 \cdot 10^{17}\ cm^{-3}$, matches the measurement in sign, shape and amplitude

simulations of the structure for voltages above the threshold voltage.

12.4.1. Sub-threshold simulations

Figures number 12.27 and 12.28 contain the results of the simulations for voltages below the threshold voltage for 1319 nm and 1064 nm laser wavelength accordingly. The well doping concentration was varied from $1 \cdot 10^{16}$ to $5 \cdot 10^{18}$ cm^{-3} (see annotations in the graphs), which in this case, is the only parameter that affects the simulations, since the oxide is described by determined properties.

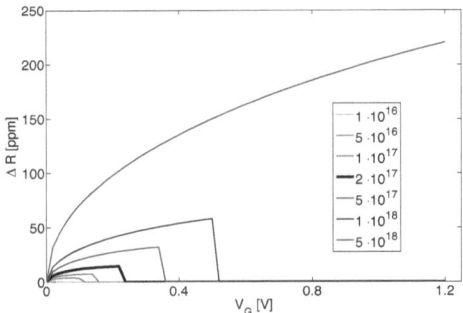

Figure 12.27.: Simulation for 1319 nm laser wavelength - reflectance at the interfaces of the varactor for voltages sub-threshold (NFET, gate), well doping concentration varied between $1 \cdot 10^{16}$ and $5 \cdot 10^{18}$ cm^{-3}

The shapes of the 1319 nm and 1064 nm wavelength simulations are similar: the absolute amplitudes increase with with increased gate voltage, until the threshold voltage is reached. The signal-to-voltage correlation is root-shaped. Since the threshold voltage calculation depends on the well doping concentration (see equation 7.7), V_{thr} is shifted to higher values as the doping concentration increases. Note that for a well doping concentration of $5 \cdot 10^{18}$ cm^{-3} V_{thr} is higher than 1.2 V, which does not make sense in terms of describing the structure. After the threshold voltage is reached, the SCR thickness, the only active signal source in this model, stays constant at the level reached and hence the amplitude of the simulation is zero. The main difference between the two sets of simulations - 1319 nm and 1064 nm wavelength - is the sign: the 1319 nm simulations predict a positive sign in opposition to the 1064 nm simulations. The amplitudes of the simulations differ as well. For the originally extracted well doping concentration (labeled black in the figures), the amplitudes of the simulations are 14.4 ppm for 1319 nm and -1274 ppm for 1064 nm wavelength.

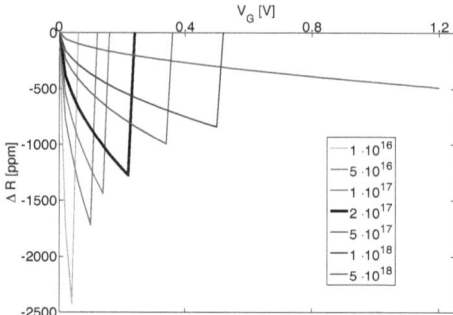

Figure 12.28.: Simulation for 1064 nm laser wavelength - reflectance at the interfaces of the varactor for voltages sub-threshold (NFET, gate), well doping concentration varied between $1 \cdot 10^{16}$ and $5 \cdot 10^{18}$ cm^{-3}

12.4.2. Simulations above threshold

Figures number 12.29 and 12.30 show the result of the simulated reflectances at the interfaces of the varactor in inversion for the NFET versus the gate voltage for the two wavelengths. For the simulations of the varactor in inversion, the influencing parameters of the active signal are the SCR and the inversion channel charge carrier density, which are both influenced by the well doping concentration (influences the threshold voltage and the SCR thickness). For both wavelengths, the well doping concentration was varied between $1 \cdot 10^{16}$ and $5 \cdot 10^{18}$ cm^{-3}, which had little effect on the amplitude and the shape of the simulation. However, the simulations for 1319 nm and 1064 nm differ in amplitude for the same well doping concentration: for $2 \cdot 10^{17}$ cm^{-3} the 1319 nm simulation had an amplitude of 115.8 ppm at 1.2 V, whereas the 1064 nm simulation had an amplitude of 43.96 ppm at the same voltage. Figure 12.29 shows three simulation results as examples for a variation in well doping concentrations (1319 nm wavelength): for all cases, the simulation values stay at zero ppm, until the threshold voltage is reached (for simulations of sub-threshold voltages see section 12.4.1). After reaching the threshold voltage, the amplitude of the simulation increases slightly until a voltage of 0.8 V is reached, then the amplitude increases rapidly. The amplitudes of the simulations at 1.2 V for $2 \cdot 10^{17}$ and $1 \cdot 10^{18}$ cm^{-3} well doping concentration vary only slightly: 115.8 ppm and 117.5 ppm. The main difference between the two simulations is the threshold voltage, after which the amplitude of the simulation increases. The estimated threshold voltage for $5 \cdot 10^{18}$ cm^{-3} is 1.2023 V, this is the reason, why the values of the simulation are zero for all voltages up to 1.2 V (compare to figure number 12.27; for calculations of the threshold voltage see section 7.4.2). To investigate the influence of the SCR thickness on the reflectance even further, the SCR thickness was varied by adding factors to it, but up to a factor of 10, the amplitude changes only slightly (results not shown in the graphs). The remaining influencing parameter is the inversion channel charge carrier density, which was calculated from equation 7.5. The formula was extracted from the electrical data simulation and hence might not represent the real inversion channel charge carrier density too well: the values could be higher or lower. To investigate the influence of the carrier density, it was multiplied by another factor. Two examples of the simulations can be found in figure 12.30 (for 1064 nm). An increase in the inversion channel carrier density by a factor of two increased the amplitude of the simulation by a factor of around four for both wavelengths. All results of the simulations were positive.

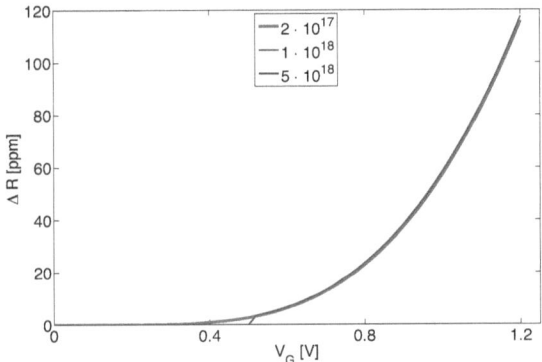

Figure 12.29.: Simulation for 1319 nm laser wavelength - reflectance at the interfaces of the varactor in inversion (NFET, gate), well doping concentration varied: $2 \cdot 10^{17}$, $1 \cdot 10^{18}$, $5 \cdot 10^{18}$ cm^{-3}

12.4.3. Summary

The simulations for voltages above the threshold voltage of both wavelengths produce very similar results: they are positive, and show a steep slope close to the nominal voltage. The 1319 nm wavelength simulations show a factor of 2.6 higher levels than the 1064 nm wavelength. Variations of the well doping concentration and adding a factor to the SCR width had only little effect on the amplitude of the simulation. A large effect made the factor for the inversion carrier density: the higher the inversion charge carrier density, the higher the amplitude. An increase by a factor of two increased the amplitude by little more than a factor of 4 for both wavelengths.

The two simulations, which match the measured signal amplitudes (1319 nm measure-

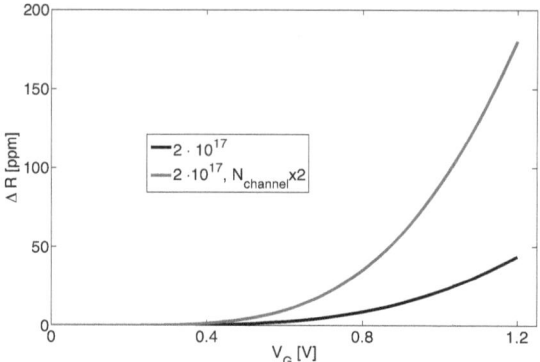

Figure 12.30.: Simulation for 1064 nm laser wavelength - reflectance at the interfaces of the varactor in inversion (NFET, gate), well doping concentration set to $2 \cdot 10^{17}$ cm^{-3}, inversion channel charge carrier density calculated and multiplied by a factor of two

ments: around 110 ppm, 1064 nm measurement: around 170 ppm, compare to 9.2.1) the best are shown in the graphs 12.29 and 12.30, too: for both wavelengths the well doping concentration was set to $2 \cdot 10^{17}$ cm^{-3}, for the 1064 nm simulation the channel carrier density was multiplied by a factor of two in order to reach the amplitude. The resulting maximum amplitudes of the simulations are 116 ppm (1319 nm) and 180 ppm (1064 nm).

The measured gate signals for the NF-L-120 had a "hump" for voltages below 0.5 V (1319 nm) and 0.3 V (1064 nm). At the nominal device voltage, the sign of the measurement with the 1319 nm laser was negative, whereas the sign for 1064 nm was positive. It can be assumed that interference is capable of changing the sign and the amplitude of a signal. It is even possible that interference reduces the very strong subthreshold simulation levels (-1274 ppm for 1064 nm wavelength, see section 12.4.1) to the signal levels, which have been measured at those voltages. So with these effects (changing sign and amplitude), it is possible to explain the according measurement results - including the "hump" - simply by interference of the simulation results shown in section 12.4.1 and above.

12.5. Simulation results: PFET, gate

As explained in section 12.4, also for the PFET gate, the simulations of the reflectance at the interfaces of the varactor will proceeded in two steps - sub-threshold simulations and simulations for voltages above the threshold voltage.

12.5.1. Sub-threshold simulations

The results of the simulations below threshold are identical - in amplitude, sign, shape and effect of the well doping concentration - to those shown in section 12.4.1 and hence are not repeated here.

12.5.2. Simulations above threshold

Figures number 12.31 and 12.32 contain the result of the simulated reflectances at the interfaces of the varactor in inversion for the PFET versus the gate voltage for the two wavelengths.

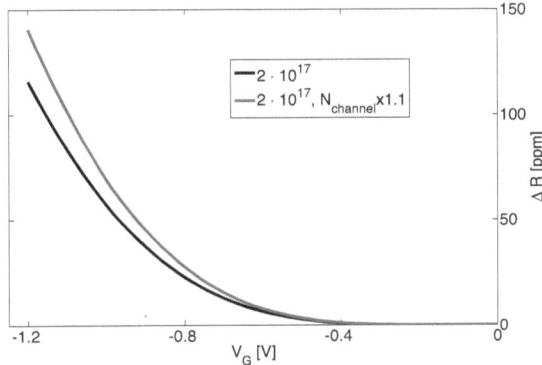

Figure 12.31.: Simulation for 1319 nm laser wavelength - reflectance at the interfaces of the varactor in inversion (PFET, gate), well doping concentration set to $2 \cdot 10^{17}$, inversion channel charge carrier density calculated and multiplied by a factor of 1.1

The simulations of the PFET varactor are almost identical to those of the NFET (see section 12.4 above): the well doping concentration and SCR thickness barely influenced the amplitude of the simulation, the signal-to-voltage correlation is the same for both wavelengths and as above, the amplitude of the simulation for 1319 nm is larger than the result of the 1064 nm. Even the multiplication factor in the channel charge carrier density affects the amplitude in the same way.

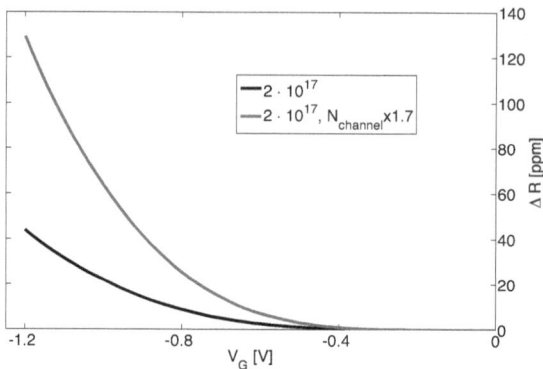

Figure 12.32.: Simulation for 1064 nm laser wavelength - reflectance at the interfaces of the varactor in inversion (PFET, gate), well doping concentration set to $2 \cdot 10^{17}$ cm^{-3}, inversion channel charge carrier density calculated and multiplied by a factor of 1.7

12.5.3. Summary

With the simulations above, the overall signal amplitude at the nominal device voltage of the measured signals (1319 nm measurement: around 130 ppm, 1064 nm measurement: around 122 ppm, see section 9.2.1) can be reproduced by choosing the factor for the inversion channel charge carrier density. The results that match the measured signal amplitudes the best are also shown in the figures number 12.31 and 12.32. The well doping concentrations were set to $2 \cdot 10^{17}$ cm^{-3} and the factors for the inversion channel carrier density were set to 1.1 for the 1319 nm simulation (amplitude 141 ppm) and to 1.7 for the 1064 nm simulation (amplitude 130).

The PFET gate signal measured with the 1319 nm laser had a linear signal-to-voltage correlation and a negative sign. The measurement with the 1064 nm laser showed a signal that had a "hump" at lower voltage levels and had a negative sign at the nominal voltage. As assumed above, it is likely that the resulting signals of sub-threshold simulations and simulations for voltages above the threshold voltage interfere with different sign and amplitude, such that the measured signals can be reproduced, as well.

12.6. Discussion and summary of the results

12.6.1. Reverse biased diodes

The simulations of the reflectance from the interfaces of the reverse biased diode, NFET *and* PFET drains, produced a broad spectrum of results. The results vary in sign, amplitude and shape over a wide range - from -600 ppm up to +40 ppm and from linear to parabolic signal-to-voltage correlations -, depending on the well and diffusion doping concentrations used for the modeling. The influencing parameters were extracted. The investigation concluded that the SCR thickness influences mainly the shape of the simulation, the refractive index and the absorption coefficient change the amplitude of the simulation, whereas the latter had not such a strong effect. Other parameters (like the mobility and the built-in potential, which also vary with the doping levels) played a minor role. The simulations, which matched the measurements the best, were performed with the following data sets:

- NFET, 1319 nm: diffusion doping concentration $2 \cdot 10^{21}$ and well doping concentration $4 \cdot 10^{18}$ cm^{-3},

- NFET, 1064 nm: diffusion doping concentration $2 \cdot 10^{21}$ and well doping concentration $1 \cdot 10^{18}$ cm^{-3},

- PFET, 1319 nm: diffusion doping concentration $4 \cdot 10^{21}$ and well doping concentration $2 \cdot 10^{19}$ cm^{-3},

- PFET, 1064 nm: diffusion doping concentration $2 \cdot 10^{21}$ and well doping concentration $2 \cdot 10^{17}$ cm^{-3}.

The resulting simulations consisted of linear signal-to-voltage correlations with various levels and signs (for details see sections 12.2 and 13.3). Except the well doping concentration of the 1064 nm simulation for the PFET drain, all doping concentrations - for well and diffusion - are well above the original values (well: $2 \cdot 10^{17}$ and diffusion: $2 \cdot 10^{20}$ cm^{-3}).

12.6.2. Varactors in inversion

The simulations of the gates were proceeding in two steps: a sub-threshold simulation and a simulation for voltages above the threshold voltage.

The simulations of N- and PFET were the same. The signal-to-voltage correlations of the sub-threshold simulations were identical for the two wavelengths: the absolute amplitude increased root-shaped up to the threshold voltage, then stayed at zero, because the SCR thickness is constant for higher voltages. The sign of the 1319 nm simulations was positive, whereas the 1064 nm simulations produced a negative sign. For the originally extracted well doping concentration, the amplitudes were 14.4 ppm for 1319 nm and -1274 ppm for 1064 nm wavelength.

All simulations for voltages above the threshold voltage were very similar. The results had a positive sign and showed a steep slope close to the nominal device voltage. Changes in the well doping concentration and SCR thickness made little effect on the amplitude of the simulation, but adding a factor for the inversion carrier density increased the amplitude of the simulation drastically. The best match between the amplitudes of the simulations and the measurements was achieved with the following factors for the inversion charge carrier densities (all well doping concentrations $2 \cdot 10^{17}\ cm^{-3}$, because it barely influenced the simulation):

- NFET, 1319 nm: no factor needed,
- NFET, 1064 nm: factor 2,
- PFET, 1319 nm: factor 1.1,
- PFET, 1064 nm: factor 1.7.

All results had the same signal-to-voltage correlation, but different amplitudes (for details see section 12.4 and 12.5).

The composition of sub-threshold simulations and simulations above V_{thr} cover the gate voltage range completely.

It is important to note that the calculated reflectances only predict the **active** signal contributions with their according amplitudes, signs and signal-to-voltage correlations and do not determine the exact overall reflectance of the device under test (see discussions about the waves that interfere within the structure and at the detector, chapter 10 and 11). It was assumed that interference of signals - taking passive signal contribution into account - is likely to cause signals of shapes and amplitudes as found in the according measurements (various signal levels, "hump"-shaped signals as well as linear signal-to-voltage correlations).

12.6.3. Limits of the Modeling

Beside the fact that the simulation results partly match the measurements well, the limitations of the modeling still should be considered. The strongest limitation of the model lies within the approximations of the structure being evaluated: the doping profiles were approximated by using average doping concentrations, so the influence of a profile-shaped "layer" is not clear. Determination of a profile as a simple layer is questionable anyway. The SCR were assumed to be fully depleted, which is not the case in a real device. And the SCR thickness and the other electrical parameters (section 7.3) were calculated from simple formulas due to the lack of "real" data (even the electrical simulation data would not help, because also these values are calculated and do not exactly represent the real situation). Other ways of calculating those parameters might influence the calculations of the reflectance. Note further that, for the calculations of the

1064 nm laser wavelength, electron-hole pair generation has been neglected. All these factors could influence the calculations of the overall reflectance and thus produce an even more diverse spectrum of simulations.

The simplifications might as well be a reason, why e.g. results that matched the measurements the best were achieved by using charge carrier densities beyond the values that were originally assumed (e.g. diffusion doping concentrations as high as $4 \cdot 10^{21}$ cm^{-3}).

Part VII.

Future prospects, summary and conclusions

13. Future Prospects

As outlined in the introduction, failure analysis tools can become obsolete quickly, if they are not capable of adjusting to the needs of future technologies and scaling. It can be assumed that failure analysis in the near future will mostly deal with SOI(silicon on insulator)-devices and process technologies of 32 nm feature size. This chapter will discuss the abilities of the LVP tool used in terms of these two future prospects.

13.1. Scaling of devices - future process technologies

Future process technologies will introduce feature sizes of 32 nm and less. This means that the nominal gate dimensions will decrease by a factor of four in comparison to the nominal test structures chosen for this work (the measurements were performed with the nominal devices of the 120 nm technology). The measurements in sections 9.2.3 and 9.3.3 revealed that strong signals in the gate area of the nominal devices were detected with the modulation amplitude maps, when both pins, gate and drain, were pulsed, otherwise the signals in the gate area were barely measurable. The signal strengths were comparable to, or even higher than, the ones of the according 10 μm structures. This indicates that the two active signal sources of gate and drain interfere, due to the small area of origin, and both, enhanced by the interference, contribute to the overall signal (the extracted signal-to-voltage correlations of the nominal devices indicate this as well, because the shape of the curves seemed to be some kind of a mixture of gate and drain signals). It is likely that signals of even smaller devices follow the same signal contribution pattern and hence will still be measurable. It might even be possible that the signal level increases with the decreased feature size, because in addition to the interference effects that get stronger, technology parameters - as e.g. higher channel doping - might increase the signal strength (as an example for increasing signal levels due to higher inversion charge carrier densities see section 12.4.2). Further, future process technologies might be built in SOI technology, which will be discussed in the following section. However, it is important to note, that signal extraction from the nominal structures was possible, because MAM and MSM methods were used. Probing the nominal gates was a challenging task and not always successful. This suggests mapping methods as the analysis tools of choice for future technologies.

13.2. SOI

In an SOI device, the active areas of the FETs are built upon a buried oxide(BOX)-layer. The advantages of SOI devices include improved MOSFET scaling due to the thin body.

The sub-threshold swing is improved and the BOX-layer is a good isolation to reduce the capacitance to the substrate, which gives rise to higher speed. A disadvantage that comes along with a BOX-layer is the bad heat conduction in comparison to a bulk device [SN07].

From LVP modeling point of view, there are two main differences between a conventional bulk device and an SOI device: the BOX-layer, which represents another layer with the according properties and the fact that the development of the SCR thicknesses in the drain and gate area is changed. For quantitative estimations of LVP signals from SOI devices, a thorough investigation of the potentials, charge carrier densities and SCR thicknesses need to be done, which is beyond the scope of this work. However, in the following, basic assumptions will be discussed.

The reflections off the BOX-layer are assumed to be large, which results in a strong static signal that influences the imaging of the structure. If the reflections get too high, the image will saturate (for the setup described in this work, the image would be completely white). This can be avoided by reducing the incident power of the laser beam or by simply adjusting the detection scheme the way that the static reflectance is suppressed. As long as the modulation levels in the reflected light are still strong enough to be detected, LVP measurements of SOI devices were still possible.

In addition, the BOX-layer causes changes in the modulations of the SCRs: since the layer limits the thickness of the active areas, the SCRs are not longer capable of expanding into the well, as it was the case for the bulk devices. This means that, for the SCR underneath the gate, the SCR thickness is limited. In the diffusion area - if the SCR thickness is modulated at all - the SCR can only expand into the diffusion, so only into the region of higher doping concentration, which results in smaller SCR thicknesses.

For the gate signals, which were modeled in section 12.1.2 (for simulation results see sections 12.4 and 12.5), this meant that the signals at gate voltages *below* threshold - related to the modulation of the SCR thickness - the signal strength is likely to be distinct from the levels of the bulk device. However, it is not clear, whether the signal will increase (smaller variations in the SCR thickness do not necessarily cause lower signal levels, see discussion below) or decrease; it might as well be zero, if the signal source, SCR modulation, is not existing for SOI devices. The signal level for voltages *above* threshold could increase, if the charge carrier densities in the channel increase (e.g. compare to section 12.4.2).

Smaller SCR thickness variations do not necessarily result in lower signal amplitudes. This can be found in the simulations of the diodes in reverse bias in sections 12.2 and 12.3: the simulations contain results of a SCR, calculated from $2 \cdot 10^{17}$ cm^{-3} well doping concentration and $2 \cdot 10^{20}$ cm^{-3} diffusion doping concentration ($\Delta t_{SCR} = 42$ nm) and $7 \cdot 10^{18}$ cm^{-3} well doping concentration and $2 \cdot 10^{21}$ cm^{-3} diffusion doping concentration ($\Delta t_{SCR} = 7$ nm) and the simulations predict signal levels of around 0.2 ppm and 40 ppm for 1319 nm respectively. Note that for SOI devices, the well is non-existing, such that the SCR developments might be completely different from those of a conventional bulk device. The SCRs could expand *horizontally* into the gate region, which might as well change the signal contribution in that particular area.

In addition, interference effects might play an even more important role in the overall signal generation of SOI devices, due to the high reflectivity of the BOX-layer. Interference could cause an increase or a decrease of the signal strengths. The worse heat conduction in comparison to a bulk device might also increase the probability that thermo-optic effects occur, as discussed in section 4.3.1, which in turn could cause higher signal levels. The drastic changes in the layout of an SOI device, with all the changes in the distributions and strengths of the fields, might lead to an even more complex system of effects generating the signal (e.g. effects, depending on the electric field, see section 4.4).

14. Summary and conclusions

This work used a modified LVP setup, which employed a cw laser. It was possible to choose between two laser wavelengths: 1319 nm or 1064 nm. Measurements were performed for both wavelengths. The test structures from Infineon Technology AG were of 120 nm and 65 nm process technology. The sizes of the FETs were chosen to be 10 μm - in order to distinguish the different signal sources and study them separately - and devices with nominal gate dimensions - to investigate the effects on structures with decreased gate dimensions. To drive the devices, a function generator, instead of a tester as in conventional tools, was used. Three new measurement methods were introduced, all of them extracting frequency-information with a spectrum analyzer in opposition to the time-domain measurements (waveform acquisition with an oscilloscope), which are performed with the commercial LVP tools. Voltage sweeps were capable of determining the signal-to-voltage correlations. Modulation amplitude maps and modulation sign maps showed signal levels and signs of various device areas. With the low-noise frequency-domain measurement methods it was possible to acquire signals very fast: a voltage sweep takes a few seconds, modulation maps a few minutes depending on the size of the area that is to be investigated.

The modulation mapping methods enabled signal detection from an area of the IC. Areas of around 20 μm times 20 μm were scanned. On large test structures, it was possible to achieve maps of gate and drain areas separately. The signal amplitudes varied for the two wavelengths, the types of FETs, the process technologies and the sizes. It was found that device layers that seemed to be out of focus, especially metal, influenced the sign and the amplitudes of the signals. Modulation sign maps, due to the evaluation of a simple boolean information (reflection increasing or decreasing from the static value), were capable of detecting even very faint signals. Modulation mapping enabled fast detection of signal amplitudes from a small signal site. However, both methods were not able to extract signal-to-voltage correlations as fast as the voltage sweep.

Voltage sweeps were found to be a useful tool for signal-to-voltage extractions from gate and drain of the FETs. Except the devices with nominal gate dimensions, all drain signals revealed a linear signal-to-voltage correlation. The source of the signals was assumed to be the modulation of the SCR thickness. The gate signals indicated two signal sources: the SCR underneath the gate, for voltages below threshold, and the inversion channel for $V_G > V_{thr}$, which vary in width and carrier density with the applied voltage. The signal shapes varied between linear and "hump-shaped" signals for the different test structures. With the voltage sweeps, the gate areas of the nominal devices were hard to probe. The signal-to-voltage correlations that were determined eventually suggested a mixture (interference) of gate and drain signals.

The measurements of both lasers revealed signals of various signs, shapes and amplitudes. It was found that 1064 nm measurements do not necessarily result in higher signal amplitudes.

A concise model that described the interaction effects of laser light - for both wavelengths - and the device activity was built. Free carrier refraction and absorption were determined to be the main effects influencing the reflectance of a device. Simulations of drain and gate signals were performed, evaluating the free carrier effects and modulations of the SCR (drain) and SCR and the inversion channel (gate), in order to explain the signal sources and to forecast signal levels for future technologies and scaling.

The simulations of the reflectance from the interfaces of the drains revealed a broad spectrum of results. The simulations varied in sign, amplitude and signal-to-voltage correlation over a wide range, depending on the well and diffusion doping concentrations used for the modeling. The influencing parameters were extracted. The investigations concluded that the SCR thickness mainly influenced the shape of the simulation, the index of refraction and the absorption coefficient changed the amplitude of the simulation, whereas the absorption coefficient had not such a strong effect. Other parameters, like the mobility and the built-in potential, played a minor role. Simulations matching the measurements in sign, amplitude and signal-to-voltage correlation were achieved using doping concentrations well above the values extracted from the doping profiles of the actual devices (e.g. diffusion doping concentration $4 \cdot 10^{21}$ and well doping concentration $2 \cdot 10^{19}$ cm^{-3}).

The simulations of the gates proceeded in two steps. First sub-threshold simulations and then simulations for voltages above the threshold voltage were performed. In opposition to the drain simulations, the simulations of the gate signals did not show such diverse results. The signal-to-voltage correlations of the sub-threshold simulations were concluding a signal produced by the modulation of the SCR thickness. The signal-to-voltage correlations stayed the same for both laser wavelengths and variations in the well doping concentration. All simulations for voltages above the threshold voltage revealed very similar results: the simulations were of the same sign and revealed the same signal-to-voltage correlations in all cases. Changes in the well doping concentration and SCR thickness had little effect on the amplitude of the simulation, but adding a factor for the inversion carrier density increased the amplitude drastically. Simulations that matched the measurements in amplitude were found for slightly higher inversion channel charge carrier densities (multiplied by factors up to two) than estimated from the electrical simulation data set. The composition of sub-threshold simulations and simulations for voltages above V_{thr} covered the entire gate voltage range. From those simulations of the active signal source, it was possible to derive all measured signals, simply by taking interference effects into account (passive signal contribution, see above: even layers out of focus contributed to the signal).

Signal levels for future process technologies, such as smaller feature size and SOI technology, were predicted to increase in signal level rather than to decrease, due to the process parameters and the interference effects that will likely enhance the signal.

Part VIII.
Bibliography

Bibliography

[BDM+99] Mike Bruce, Greg Dabney, Shawn McBride, Jason Mulig, Victoria Bruce, Charles Bachand, Steven Kasapi, and Jefferey A. Block. Waveform acquisition from the backside of silicon using electro-optic probing. *Proceeding from the 25th International Symposium for Testing and Failure Analysis (ISTFA), Santa Clara, California*, pages 19–25, 1999.

[BDPB04] F. Beaudoin, R. Desplats, P. Perdu, and C. Boit. Principles of thermal laser stimulation techniques. *EDFAS, Microelectronics Failure Analysis, Desk Reference, 5th edition*, 2004.

[EE90] J. Eichler and H.-J. Eichler. *Laser.* Springer Verlag, 1990.

[FLM06] Bradley J. Frey, Douglas B. Leviton, and Timothy J. Madison. Temperature-dependent refractive index of silicon and germanium. *Proc. of SPIE 6273 (Orlando)*, June 2006.

[Fra58] W. Franz. Einfluss eines elektrischen Feldes auf eine optische Absorptionskante. *Z. Naturforschung 13a*, pages 484–489, 1958.

[GK95] M. A. Green and M. J. Keevers. Optical properties of intrinsic silicon at 300 k. *Progress in Photovoltaics: Research and Applications*, 3(3):189–192, 1995.

[GSG93] M. Goldstein, G. Sölkner, and E. Gornik. Heterodyne interferometer for the detection of electric and thermal signals in integrated circuits through the substrate. *Rev. Sci. Instrum. 64 (10)*, pages 3009–3013, 1993.

[HA79] P. P. Ho and R. R. Alfano. Optical kerr effect in liquids. *Phys. Rev. A*, 20(5):2170–2187, Nov 1979.

[Hec87] Hecht. *Optics.* Eddison-Wesley Publishing Company, second edition, 1987.

[Kel57] L. V. Keldysh. Behaviour of non-metallic crystals in strong electric fields. *J. Exptl. Theoret. Phys. (USSR) 33*, pages 994–1003, October 1957. translation: Soviet Physics JETP 6 (April 1958) 763–770.

[Kel58] L. V. Keldysh. The effect of a strong electric field on the optical properties of insulating crystals. *J. Exptl. Theoret. Phys. (USSR) 34*, pages 1138–1141, May 1958. translation: Soviet Physics JETP 7 (November 1958) 788-790.

[Kel64] L. V. Keldysh. Ionization in the field of a strong electromagnetic wave. *J. Exptl. Theoret. Phys. (USSR) 47*, pages 1945–1957, 1964. translation: Soviet Physics JETP 20 (1965) 1307–1314.

[KF88]   M. V. Klein and T. E. Furtak. *Optik.* Springer Verlag, 1988. translation of the American version, title "Optics", sec. ed. 1986.

[LFB+05]   M. Litzenberger, C. Fürböck, Bychikhin, D. Pogany, and E. Gornik. Scanning heterodyne interferometer setup for the time-resolved thermal and free-carrier mapping in semiconductor devices. *IEEE Transactions on Instrumentation and Measurement*, 54(6):2438–2445, 2005.

[LNB+04]   William Lo, Nagamani Nataraj, Nina Boiadjieva, Praveen Vedegarbha, and Kenneth Wilsher. Polarization difference probing: A new phase detection scheme for laser voltage probing. *Proceedings from the 30th International Symposium for Testing and Failure Analysis (ISTFA), Massachusetts*, 2004.

[MBE73]   T. S. Moss, G. J. Burrell, and B. Ellis. *Semiconductor Opto-Electronics.* Butterworth & Co. Ltd., first edition, 1973.

[Pan71]   J. I. Pankove. Optical processes in semiconductors. *Dover Publications Inc. New York*, 1971.

[SB87]   R. A. Soref and B. R. Bennett. Electrooptical effects in silicon. *IEEE Journal of Quantum Electronics*, QE-23, 1987.

[Sie86]   A. E. Siegman. *Lasers.* University Science Books, 1986.

[SN07]   S. M. Sze and Kwok K. Ng. *Physics of Semiconductor Devices.* Wiley-Interscience, third edition, 2007.

[WC65]   P.H. Wendland and M. Chester. Electric field effects on indirect optical transitions in silicon. *Phys. Rev.*, 140(4A):1384–1390, 1965.

[WE94]   H.-G. Wagemann and H. Eschrich. *Grundlagen der photovoltaischen Energiewandlung.* B. G. Teubner Verlag, 1994.

[Wil60]   R. Williams. Electric field induced light absorption in cds. *Phys. Rev. 117*, pages 1487–1490, 1960.

[WL00]   Ken Wilsher and William Lo. Integrated circuit waveform probing using optical phase shift detection. *Proceedings from the 26th International Symposium for Testing and Failure Analysis (ISTFA), Washington*, 2000.

[Wol69]   H. F. Wolf. *Silicon Semiconductor Data.* Oxford Pergamon Press, 1969.

Further LVP related publications of the author

[KWT+07] Ulrike Kindereit, Gary Woods, Jing Tian, Uwe Kerst, Rainer Leihkauf, Christian Boit. Invited paper: Quantitative Investigation of Laser Beam Modulation in Electrically Active Devices as used in Laser Voltage Probing. *IEEE Transactions on Device and Materials Reliability, vol. 7 (1)*, pages 19–30, 2007.

[KWT+07/2] Ulrike Kindereit, Gary Woods, Jing Tian, Uwe Kerst, Christian Boit. Investigation of Laser Voltage Probing Signals in CMOS Transistors. *IEEE Proceedings of 45th IRPS Phoenix, Arizona*, pages 526–533, 2007.

[KBK+07] Ulrike Kindereit, Christian Boit, Uwe Kerst, Steven Kasapi, Radu Ispasoiu, Roy Ng, William Lo. Comparison of Laser Voltage Probing and Mapping Results in Oversized and Minimum Size Devices of 120 nm and 65 nm Technology. *Proceedings of 19th European Symposium on Reliability of Electron Devices, Failure Physics and Analysis (ESREF) Maastricht, Netherlands, Microelectronics Reliability (Elsevier), vol. 48 (8–9)*, pages 1322-1326, 2008.

[B+08] Boit et al.. Physical IC debug - backside approach and nanoscale challenge. *Advances in Radio Science (ARS), Kleinheubacher Berichte*, pages 265–272, 2008.

Part IX.
Acknowledgment

The author would like to thank the following individuals for their support during the project and the development of this work.

Special thanks go to Prof. Dr.-Ing. C. Boit, who made it possible to work in such a project in the first place. I am indebted for his encouragement throughout the project and the constructive criticism.

I wish to thank Prof. Dr. rer. nat. D. Schmitt-Landsiedel for her interest in this work, reporting and the fruitful discussions.

Thank you, Prof. Wagemann, for the advice to get rid of the TV and read good books instead in the early days of my studies and for sparking interest in semiconductor physics and devices in the first place.

Another important person for the development of this project was Gary L. Woods, who introduced me to the complex LVP setup and who never got tired of answering boring questions of the new grad student hanging around in his lab. Thank you for being a mentor as well and for cheering me up, whenever things went wrong.

I was honored to work with the employees of Credence Systems, DCG (now: DCG Systems) and I thank all of them for welcoming me so friendly, working things out together, supporting my work each day I was in California and paying for the endless expenses that enabled me to finish this work.
Special thanks go to:
William Lo (for theoretical and practical support, discussions and making sure I use my spare time for the "right" things)
Nina Boiadjieva (for making sure that I am not starving in the USA, mentally and physically)
Ken Wilsher (for discussions about high frequency stuff; the polite English gentleman was the only one, who dared to correct my English)
Jing Tian, Steven Kasapi, Ted Lundquist, Ben Cain (for support, discussions and taking over the responsibility of the project when times came)
Roy Ng, Radu Ispasoiu (for support in the lab)
The "FIB-group" members (who enabled "insight" into the devices)

All employees of Infineon Technologies AG, who were dicing and bonding devices, provided information about the test structures, showed interest in the measurements and were very supportive. Especially Peter Egger, Siegfried Görlich and Christof Brillert.

I am grateful to my colleagues (Tuba, Piotr, Arek, Taro, Philipp, Rudolf, Mahyar, Uwe) for picking me up for lunch and drinking way too many coffees, especially to the one that was discussing the world with me.
Again, and again, Rainer Leihkauf, who prepared the electrical simulations and kind of dealt with the same problems.

Werner Eschenberg, for discussing the importance of time-lines and environmental-friendly disposal of electronic devices. You made my days!

Marlies Mahnkopf, Andreas Eckert and Rene Hartmann for the preparation of the devices and Rene Hartmann, Uwe Voss and Werner Eschenberg for the support in dealing with soft- and hardware. Uwe Kerst for general support.

My karate classes - in the USA AND Germany -, my friends, my mother, my grandmother and everybody who supported me personally, no matter if by punching the hell out of me, praying or believing in me as if I as well could become an astronaut! And especially my girlfriend Heike for cheering me up all the time and pushing away the dark clouds as far as possible. Kiai!

Part X.
Appendix

14.1. Free carrier absorption and refraction

The classical theory of dispersion in dielectrics predicts that the optical constants in dielectrics depend on the wavelength and the frequency (compare to [MBE73]). The explanation for this effect can be obtained from the simple classical treatment of Lorentz, which considers the solid as an assembly of oscillators, which are set into forced vibration by the radiation. The equation of motion of an electron can be written as in 14.1.

$$m_0 \frac{d^2 x}{dt^2} + m_0 g \frac{dx}{dt} + m_0 \omega_0^2 x = -q E_x e^{i\omega t}. \tag{14.1}$$

Wherein $E_x e^{i\omega t}$ is the applied electric field (incident electromagnetic wave of radiation), x is the displacement of the electron, $m_0 \omega_0^2 x$ is the restoring force, $m_0 g \frac{dx}{dt}$ is the damping, with the angular frequencies ω_0 and g (g is called damping factor) and m_0 the mass of the electron in vacuum.

A solution of this equation shows that x varies sinusoidally at the applied frequency with the complex amplitude given by

$$x_0 = \frac{\frac{-qE_x}{m_0}}{\omega_0^2 - \omega^2 + i\omega g}. \tag{14.2}$$

With equation 14.2 the dielectric constant can be derived from:

$$D = \epsilon E = \epsilon_0 \epsilon_* E + P, \tag{14.3}$$

with $\epsilon_* = 1$ when dealing with the fundamental electronic absorption band and $\epsilon_* = n_0^2$ in the infra-red absorption band. With the polarization $P = Nqs$ this equation becomes:

$$\epsilon_0 \epsilon_r E = \epsilon_0 \epsilon_* E + Nqs, \tag{14.4}$$

and in a one-dimensional form, for a displacement of the electron against field direction (i. e. negative sign):

$$\epsilon_r = \epsilon_* - \frac{Nqx_0}{\epsilon_0 E_x}. \tag{14.5}$$

With equation number 14.2 the dielectric constant becomes:

$$\epsilon_r = \epsilon_* + \frac{Nq^2 E_x}{m\epsilon_0 E_x (\omega_0^2 - \omega^2 + i\omega g)} \equiv (n - ik)^2 \tag{14.6}$$

$$\epsilon_r = n^2 - k^2 - i2nk = \epsilon_* + \frac{Nq^2}{m_0 \epsilon_0 (\omega_0^2 - \omega^2 + i\omega g)}. \tag{14.7}$$

Separation of the real and imaginary part gives:

$$n^2 - k^2 - \epsilon_* = \frac{Nq^2}{m_0\epsilon_0} \cdot \frac{(\omega_0^2 - \omega^2)}{(\omega_0^2 - \omega^2)^2 + (\omega g)^2} \qquad (14.8)$$

and

$$2nk = \frac{Nq^2}{m_0\epsilon_0} \cdot \frac{\omega g}{(\omega_0^2 - \omega^2)^2 + (\omega g)^2}. \qquad (14.9)$$

For absorption due to free carriers, there is no restoring force and hence $\omega_0 = 0$; and since the electrons are not completely free as in vacuum, the mass must be replaces the the effective mass m. Equations 14.8 and 14.9 thus become:

$$n^2 - k^2 - \epsilon_* = -\frac{Nq^2}{m\epsilon_0} \cdot \frac{\omega^2}{(-\omega^2)^2 + (\omega g)^2} \qquad (14.10)$$

and

$$2nk = \frac{Nq^2}{m\epsilon_0} \cdot \frac{\omega g}{(-\omega^2)^2 + (\omega g)^2}. \qquad (14.11)$$

With τ, the collision time, the damping factor can be written as $g = \tau^{-1} = \frac{q}{\mu m}$ (μ being the mobility). The plasma frequency is defined as $\omega_p = \sqrt{\frac{Nq^2}{\epsilon_0 m}}$. With the simplification $\tau^2 \omega^2 \gg 1$, the equations can be written as:

$$n^2 - k^2 - \epsilon_* = -\frac{\omega_p^2}{\omega^2} \qquad (14.12)$$

and

$$2nk = \frac{\omega_p^2 g}{\omega^3}. \qquad (14.13)$$

From equation number 14.12, the change in the index of refraction (Δn) due to free carriers (ΔN) is calculable as follows (by concluding that $n^2 \gg k^2$ and $\omega = \frac{2\pi c_0}{\lambda}$):

$$\Delta n = \sqrt{n_0^2 - \frac{(q\lambda)^2}{4(\pi c_0)^2 \epsilon_0} \cdot \frac{\Delta N}{m}}, \qquad (14.14)$$

for the infra-red absorption band (our case!)

$$\Delta n = \sqrt{1 - \frac{(q\lambda)^2}{4(\pi c_0)^2 \epsilon_0} \cdot \frac{\Delta N}{m}} \qquad (14.15)$$

for the fundamental absorption band. And with $n = n_0 + \Delta n$ - here, n_0 is the index of refraction with taking dispersion into account (compare to section 4.2.2) - and Δn from equations 14.14 or 14.15 and equation number 1.6, the change in the absorption coefficient due to free carriers is then calculable, too:

$$\Delta\alpha = \frac{\lambda^2 q^3}{4\pi^2 c_0^3 \epsilon_0} \cdot \frac{\Delta N}{nm^2 \mu}. \qquad (14.16)$$

In the presence of both charge carrier types, electrons and holes, ΔN becomes $\Delta N_e + \Delta N_h$.

14.2. Calculation of the mobility

Wolf describes the mobility μ for holes and electrons in silicon depending on

- the lattice of the crystal - $\mu_{lattice}$
- the impurity concentration (doping concentration) - $\mu_{impurity}$ and
- the density of free carriers (electrons or holes) - $\mu_{freeCarrier}$.

The overall mobility is calculable as follows (see formula in [Wol69], section 2.3):

$$\frac{1}{\mu} = \frac{1}{\mu_{impurity}} + \frac{1}{\mu_{lattice}} + \frac{1}{\mu_{freeCarrier}}. \qquad (14.17)$$

Wherein

$\mu_{lattice,e} = 2.1 \cdot 10^9 \cdot T^{-2.5}\ [\frac{cm^2}{Vs}]$,

$\mu_{lattice,h} = 2.3 \cdot 10^9 \cdot T^{-2.7}\ [\frac{cm^2}{Vs}]$,

$\mu_{impure} = \dfrac{2^{3.5} \cdot (\epsilon_{Si}\epsilon_0)^2 (kT)^{1.5}}{\pi^{1.5} N_{impure} q^3 m^{0.5} ln\left[1+(3\epsilon_{Si}\epsilon_0 kT/q^2 N_{impure}^{1/3})^2\right]}$ and

$\mu_{freeCarrier} = (m_e^{-1} + m_h^{-1})^{0.5} \dfrac{3(\epsilon_{Si}\epsilon_0)^2 (kT)^{1.5}}{2^{1.5}\pi^{0.5} q^3 N_{freeCarrier}} \dfrac{1}{ln\left[1+4\epsilon_{Si}\epsilon_0 kT/q^2 N_{freeCarrier}^{1/3}\right]^2}.$

With N_{impure} the doping concentration and $N_{freeCarrier}$ the free carrier concentration in $[cm^{-3}]$.

There are four cases, for which the mobility needs to be calculated separately:

- electrons in n-type Si,
- electrons in p-type Si,
- holes in p-type Si and
- holes in n-type Si.

For the calculations of these cases the according free carrier and doping concentrations and the masses need to be taken into account.

To determine the mobility at very high doping concentrations, graph 2.3.1 in [Wol69] was used. The according mobilities for electrons and holes are 140 (electrons) and 100 (holes) $\frac{cm^2}{Vs}$.

14.3. Setup components - tool specification

The setup described in part III employed various components. The following sections specify the most important equipment used for this work.

14.3.1. Oscilloscope

LeCroy
SDA 6000A XXL (serial data analyzer) [1]
Analog Bandwidth at 50 Ω (-3 dB): 6 GHz
Rise Time (Typical): 75 ps
Input Channels: 4
Input Impedance: 50 Ω ±2.0 percent
Maximum Input Voltage: ±4 V_{peak}
Sensitivity: 2 mV to 1 V/div (fully variable, < 10 mV/div through zoom)
Time/Division Range: Real Time: 20 ps/div to 10 s/div
Single-Shot Sample Rate/Channel: 20 GS/s on 2 Ch.; 10 GS/s on 4 Ch.
Averaging: Summed averaging to 1 million sweeps; continuous averaging to 1 million sweeps
Front Panel and Instrument Status: Store to the internal hard drive or to a USB-connected peripheral device.

14.3.2. Spectrum analyzer

Advantest
AdvantestR3162 [2]
Low end frequency limit: 9.00 kHz
High end frequency limit: 8.00 GHz
Resolution Bandwidth Min.: 1 kHz
Resolution Bandwidth Max.: 3 MHz
Amplitude Range: +30 dBm to average noise level
Total Level Accuracy: Maximum 1.5 dB
High-speed GPIB, Effective for System Applications

---

[1] a more detailed specification can be found here: http://www.lecroy.com/tm/products/analyzers/sda/default.asp
[2] a more detailed specification can be found here: http://www.testbuyer.com/pdf/specs.cfm?pdf$_i$d = 550F570CBA

High Speed Measurement: 20 traces per second
6.5-inch TFT color LCD
Built-in Frequency Counter
Printer Port
20 dB Preamp

14.3.3. Function generator

Tektronix
Tektronix HFS9003 (Programmable Stimulus System) [3]
Fully programmable
Number of Channels: 12
Frequency Range: 50 kHz to 630 MHz
Phase lock mode
Outputs:
Maximum HIGH level: +5.00 V
Minimum LOW level: -2.50 V
Accuracy: 2 percent
Impedance: 50 Ω
Trigger Input Performance:
Input Resistance: 50 Ω
Input Voltage Range: ± 5 V maximum
Ports to Peripheral Devices: GPIB
Out of Production: Jul-11-2005

---

[3] a more detailed specification can be found here: http://www.testmart.com/webdata/mfr$_p$$df s/TEK/TEK_H FS.pdf$

i want morebooks!

Buy your books fast and straightforward online - at one of world's fastest growing online book stores! Environmentally sound due to Print-on-Demand technologies.

Buy your books online at
www.get-morebooks.com

Kaufen Sie Ihre Bücher schnell und unkompliziert online – auf einer der am schnellsten wachsenden Buchhandelsplattformen weltweit! Dank Print-On-Demand umwelt- und ressourcenschonend produziert.

Bücher schneller online kaufen
www.morebooks.de

 VDM Verlagsservicegesellschaft mbH
Heinrich-Böcking-Str. 6-8   Telefon: +49 681 3720 174   info@vdm-vsg.de
D - 66121 Saarbrücken   Telefax: +49 681 3720 1749   www.vdm-vsg.de

Printed by Books on Demand GmbH, Norderstedt / Germany